Preface to the third edition

Since the first edition of this book was published enormous progress has been made in many fields of biology, not least in microbiology and plant physiology. School courses have tended to emphasize the need for good practical work and proper experimental design and the study of plant physiology has come to occupy a vital position particularly as physiological work with animals is now often regarded as undesirable or too expensive to carry out. In this edition a number of sections have been modified and brought up to date and considerable new material added on the topics of plant growth and development. Rather recently there has been some trend towards more study of the whole plant, particularly in relation to its environment, and in the final chapter, in particular, an attempt has been made to think of the life-cycle of the whole plant. On the other hand the earlier chapters continue to bring out ways in which the plant interacts with its environment and so is able to compete successfully or unsuccessfully with its neighbours.

It is often felt that each step in a scientific inquiry should be based on practical results. This is clearly an ideal that cannot always be met; nevertheless, practical work has been included in almost every section. Some of the practical techniques have been slightly modified, with more instructions over the care required in handling the various materials. Some new techniques are described and one or two of the more complex procedures have been cut out. Many of these practical techniques have proved of great value in project work, but the student must not treat them as so many recipes; they need working through carefully with properly designed controls and understanding of the kind of accuracy involved and of the need to carry out numbers of replicates.

This book is intended for use in advanced work in schools and also in some introductory university courses. It is hoped that it will continue to stimulate interest in physiology as well as ecology and other fields of biology.

1978   W.M.M.B.

Acknowledgments

I must express my gratitude to a number of people who have made this and earlier editions possible. Dr. V. S. Butt, Dr. B. E. Juniper and Dr. Jean Whatley of the Botany School, Oxford University have provided me with much useful information, ideas and helpful criticism as well as some superb electron microscope photographs. In the earlier editions I was greatly assisted by my former colleague Professor W. H. Dowdeswell and in this third edition I have had an enormous amount of help and encouragement from Dr. T. A. Hill and Dr. T. Peat of Wye College, University of London. They have read typescripts, provided new information and photographs through almost every section of this new edition. To them both I am truly grateful. I must also thank Dr. R. F. O. Kemp, Dr. Rachel Leach, Mrs. J. H. Preston, B. J. F. Haller of Philip Harris Ltd., the late A. S. Mitchener, Miss Grace Monger and many teachers in a variety of schools who have provided me with a vast number of useful ideas, materials and help. Finally, I should like to thank the various authors and publishers who have allowed me to make use of their materials, as well as my own publishers for their continued assistance.

Chemical Names

The names of many of the chemical substances discussed in the text have undergone changes during the last few years; the following list of synonyms may be useful.

Old name	*New name*
Acetaldehyde	Ethanal
Acetic acid	Ethanoic acid (old name still acceptable)
Acetylene	Ethyne
Citric acid	2-hydroxypropane-1,2,3-tricarboxylic acid
Ethyl alcohol	Ethanol
Ethylene	Ethene
Fumaric acid	*Trans*-butanedioic acid
Glutamic acid	2-aminopentanedioic acid
Iso-citric acid	1-hydroxypropane-1,2,3-tricarboxylic acid
α-ketoglutaric acid	1-oxobutanedioic acid
Malic acid	2-hydroxybutanedioic acid
Malonic acid	propanedioic acid
Oxaloacetic acid	2-oxobutanedioic acid
Oxalosuccinic acid	1-oxopropane-1,2,3-tricarboxylic acid
Pyruvic acid	2-oxopropanoic acid
Succinic acid	Butanedioic acid

Symbols

$\times 10^6$	Mega	M
$\times 10^3$	kilo	k
$\times 10^{-3}$	milli	m
$\times 10^{-6}$	micro	μ
$\times 10^{-9}$	nano	n

1 μm = 1 micron
1 Å = 10^{-10} m = 0.1 nm
1 cal = 4.18 J
1 atmosphere = 101.3 kPa   1 Pa = 1 Nm^{-2}

Contents

I

Introduction: Organization of Cells and Tissues

1.1 Introduction

Plant physiology is a synthesis of many aspects of botany and biology. Its study is based on structure and anatomy on the one hand, and the physical and chemical changes and organization in the whole plant and in its cells on the other. Yet these two lines of approach mean relatively little on their own; they must be related to the environment in which the plant grows and to the evolution of the species concerned. Physiology takes on a real meaning when it is realized that its various features, at a biochemical or visible structural level, may have important survival value. It is the aim of this book to show how these aspects—anatomical, physical, chemical or biochemical, genetic and ecological—combine and are vital if a plant is to survive in any environment and compete with its neighbours. By studying these factors together it is possible to obtain some understanding of the ways in which plants are organized; in short, of how they work.

1.2 Vital requirements and vital processes

Any wild plant usually occupies a clearly defined ecological niche; it is able to survive there because the site has the environmental conditions that are most favourable, providing it with its various requirements so as to enable it to compete satisfactorily with other plants. What are these vital requirements?

Perhaps the most obvious of these is the *water requirement*. All living things require water for the hydration of their protoplasm, the living material in their cells. This is largely because the various chemical and physical changes occurring need to take place in a watery medium. The ions, enzymes and organic materials are not able to move about the cell and diffuse to areas where they are required for metabolism, unless the protoplasm is well hydrated. Much of the transport of dissolved material also takes place through water movement, and water is also required in considerable quantity to maintain the turgidity of the cells and so support the plant, as well as to replace water that has been lost by evaporation. A rather smaller quantity of water is also needed in many chemical reactions, for instance in photosynthesis and in the breakdown of many large molecules by hydrolysis.

As energy is required for growth, reproduction and the majority of changes going on within the plant, it is vital to possess in the first place a mechanism for obtaining and storing energy, and in the second place, a means for releasing this stored energy for use in the various metabolic processes going on in the plant. Green plants obtain their energy through the process of *photosynthesis*, light being

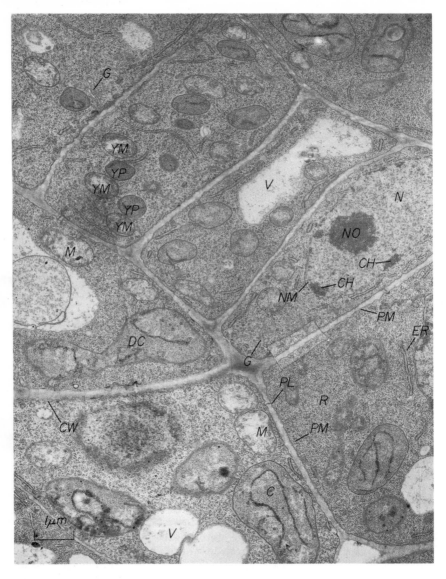

Fig. 1.1. Electron microscope photograph of cells from the young leaf of the runner bean (*Phaseolus*). One nucleus (N) shows its nucleolus (NO) particularly clearly as well as patches of darker chromatin (CH). The nuclear membrane (NM) is also clear. The cytoplasm consists of a mass of free ribosomes (R) together with various organelles. There are young plastids (YP) and young mitochondria (YM) and a small vacuole (V), as well as older chloroplasts (C), one of which is dividing (DC) and older mitochondria (M). There are several patches of endoplasmic reticulum (ER) and a number of Golgi apparatus are also clear (G). Each cell is bounded first by a plasma membrane (PM) which appears as a thin dark line and then by the cell wall (CW) of cellulose which is pierced by a number of plasmodesmata (PL) which may afford communication between the cells. (Courtesy of Dr. Jean Whatley.)

absorbed by the green *chlorophyll* in the *chloroplasts* (see fig. 1.1); *carbon dioxide* is also required and is taken up from the atmosphere. The energy is stored principally in the form of sugars and other carbohydrates. Not all plants have this *light requirement*; some are non-green and must depend on the organic materials made by green plants for their energy sources; others, the *chemosynthetic* organisms, are totally independent of the sun's energy and use energy released from inorganic changes.

Respiration is the vital process by which carbohydrates and other complex molecules are broken down. This process is usually one of oxidation requiring atmospheric oxygen and yielding energy, which can then be used for the various metabolic processes going on in the plant.

Finally, although plants can obtain the elements that they require for photosynthesis from water in the soil and from carbon dioxide in the air, nevertheless, they need many other elements. For instance, nitrogen is required for the vitally important *proteins*, which, through the organic catalysts, the *enzymes*, are one of the most important groups of organic materials in living things. Nitrogen is seldom utilized by plants in the gaseous state; most of it is taken up from the soil solution in the form of simple inorganic ions, such as nitrate. Other elements such as phosphorus, sulphur, magnesium, copper and iron, are also required for elaboration into various organic molecules, and these too are taken up in the form of inorganic ions; these substances are referred to as the *mineral nutrients*. The fertility of the soil depends to a great extent on the availability of these substances; plant distribution too is closely related to mineral availability, as different plants are differently adapted, and some can tolerate more or less of one mineral than another. Consequently mineral nutrition is another important aspect of physiology.

In this book it is intended to show, first, how and where the vital processes are carried out, particularly within the organization of the basic unit of construction, namely *the cell*, and secondly, something of how these processes are integrated, both one with another and relative to the organization of the plant as a whole.

1.3 The cell as the basic unit

One of the most obvious and in a sense unifying features of plant and animal structure is that the vast majority are composed of cells. True, there are exceptions, notably the slime moulds (*Myxomycetes*) and the coenocytic algae and fungi, which contain many nuclei but no distinct cellular system. But most plants are unicellular or multicellular, and since the time of Schwann in 1839, it has been realized that there must be something about cells which makes them a necessity to the more highly adapted organisms. In higher plants it seems likely that the cell represents the smallest organized mass of material that is capable of fending for itself, of living. Why is this so? The answer to some extent lies in the structure of the cell itself, and for the rest in the organization of the vital processes that go on inside it.

Although there is a great variety of cells, each specialized for a particular function, a simple, relatively unspecialized cell, such as found in the growing region of the root apex (see fig. 1.2A), is bounded by a thin, elastic cell wall. Half of this is composed of the fibrous polysaccharide, cellulose, the rest being mainly packing of hemicellulose polysaccharides and various pectates. This encloses a granular, heterogeneous, watery mass consisting largely of protein, called the *cytoplasm*. Floating in the cytoplasm is the rounded *nucleus*, the controlling unit of the cell. Apart from water, both the cytoplasm and nucleus are composed largely of protein, and are referred to jointly as the *protoplasm*.

A cell is usually unable to survive for long without its nucleus. If the nucleus is removed with a micro-manipulator the cell will soon die; if it is replaced it will continue to live. An exception to this is the phloem sieve tube which contains no nucleus though it is likely that the control of the cell is vested to some extent at least in the nucleus of the neighbouring companion cell (see fig. 1.6). The nucleus is known to contain the hereditary factors or *genes* which control the organisation of the cell; these genes produce messenger substances which are capable of directing the synthesis of proteins and enzymes and these determine the patterns of cell physiology.

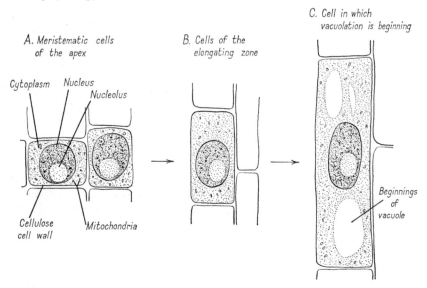

Fig. 1.2. Differentiation in the root tip of the broad bean. (×1000.) The cells shown would all develop into the much elongated xylem elements.

Observation of the cell under the phase-contrast microscope (which makes the cell structures much clearer without killing the cell and using stains) shows the whole to be in a state of constant motion. This is due to the small particles in the cytoplasm jostling one another partly as a result of the thermal activity of the dissolved ions and molecules hitting the larger particles, a phenomenon known as *Brownian movement*, and partly as a result of metabolically controlled activity called protoplasmic streaming. Careful examination of the granular cytoplasm shows the presence of minute ellipsoid structures called *mitochondria*. These are the sites of respiration and tend to congregate in areas where energy is required. Electron microscope photographs of thin sections of cells show many smaller structures in the cytoplasm (see fig. 1.1) which may add to the granular appearance of the cell when viewed under an ordinary light microscope. These smaller structures, the endoplasmic reticulum, ribosomes and Golgi bodies, can be grouped together as the *microsomal* complement of the cell and are the areas where proteins are synthesized (see p. 131).

1.4 Cell development

During growth there are essentially three phases of cell development; first *cell division* takes place, second there is a phase of *extension* or elongation of the cell in which the rate of division is greatly reduced. Finally *cell differentiation* becomes apparent as the cells become specialized for their particular functions.

Cell division

Cell division takes place most rapidly in the actively growing areas. As differentiation proceeds, the rate of division usually falls off considerably, though there may be *residual meristems*, for instance the cambium, in which division may continue throughout the life of the plant. The process of division is called *mitosis*; it reveals the presence of long, thin *chromosomes*, composed of protein and nucleic acids, that are known to be the site of the hereditary factors, the genes. Phase-contrast observation or staining techniques (see Appendix, p. 241) show these chromosomes at the onset of cell division as long, thin, double fibres, joined together at the *centromere*. During the division the centromere divides and the two daughter chromosomes are drawn to opposite sides of the cell, where they form two new nuclei. At the end of mitosis these nuclei appear much like the parent nucleus, with no distinct chromosomes visible, but before the next division takes place the cell has synthesized new chromosome material so that they are double once again. In this way mitosis provides each daughter nucleus with a full set of genetic information.

Cell extension

Immature, meristematic cells of the root tip and stem apex are more or less cubical and quite small, perhaps 10–20 μm across (see symbols p. iv). They are metabolically very active, and rapidly accumulate food materials and synthesize new protoplasm. If older cells farther back are examined (see fig. 1.2c) these will be larger and more elongated, and an additional feature, the vacuole, may be visible. The vacuole is composed mainly of water, but also contains dissolved mineral and organic materials. It acts as a reservoir for water and minerals and sometimes as a storage area for waste products of cell metabolism; it also keeps the cell inflated and turgid, and in this way, due to one cell pressing against another, helps by giving strength to tissues and support to the whole plant. Finally, the vacuole is important in cell economy. Protoplasm is expensive material; it requires energy for its synthesis and respiration to keep it alive. Water is inexpensive in such terms of cell economy, and a large cell, perhaps simply concerned in giving bulk to a part of the plant, may contain a very large vacuole though this will, of course, add considerably to the mass of the whole.

Finally, when the cell reaches its mature size and state of development it may be highly differentiated or specialized according to a particular function. An interesting point is, why is there a maximum size at all? Could not one nucleus control a great mass of protoplasm? The answer seems to be that the size of a fully developed cell (a packing or *parenchymatous* cell may be 100 μm in diameter) is dependent on the ability of the nucleus to control the work of the cell constituents. Materials diffusing out of the nucleus and giving 'instructions' for cell operation may only be capable of moving a limited distance in an organized manner. In the same way, interaction between the various cell organelles can only take place effectively over relatively short distances. The surface area of a cell is also an important factor influencing cell size. A large surface area is necessary for diffusion of substances into and out of the cell. As a cell enlarges the ratio of its surface area to volume decreases, and so cells of more than 100 μm in diameter are seldom found. Furthermore, if specialization is to take place this is an easy way to carry it out. It is simplest to start with undifferentiated building units and gradually develop these into one of a variety of specialized tissues.

Cell differentiation

Farther from the apex the cells divide less and less, usually becoming more elongated and vacuolated, gradually taking on the structure and functions of the

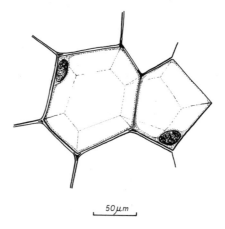

50 μm

Fig. 1.3. Parenchymatous cells from the stem cortex of the broad bean; cut-away view.

specialized cells of the more mature plant. This differentiation has always posed something of a problem to the physiologist, as it is not known how cells placed together at the apex, and usually genetically identical, are capable of developing to form totally different types of specialized cells.

It is a mistake to take any type of cell as being 'typical'; while most meristematic cells look similar to one another, they are totally different from parenchymatous, palisade and other well-differentiated types. Parenchymatous cells are specialized as packing-cells; in the mass they form the tissue *parenchyma*. They are of complex shapes and many have six-sided faces, some eight-, five, or four-sided faces. As these cells are tightly pressed together there are relatively few spaces between them and they are described as space-filling. The regular geometric shape that approaches nearest the typical parenchymatous cell is the tetrakaidecahedron (see figs. 1.3 and 1.4).

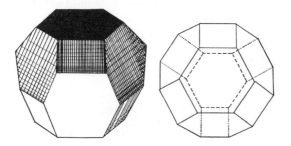

Fig. 1.4. The tetrakaidecahedron. The space-filling geometric shape that resembles most closely the typical parenchymatous cell.

The palisade cell (see fig. 1.5), on the other hand, is specialized for photosynthesis and has a far greater content of chloroplasts than most other cells in the stem and leaves. As is shown by the dense arrangement of these cells underneath the transparent epidermis, they are well adapted for this essential function.

Some of the most differentiated cells are in the conducting system. For instance, the phloem sieve tubes, which are long and narrow, filled with protoplasmic fibrils and contain no nucleus. Each is joined to the next sieve tube through the multi-perforated sieve-plate, forming a highly evolved transport

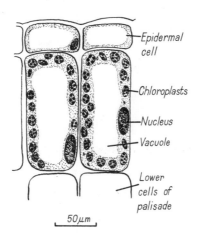

Epidermal cell

Chloroplasts

Nucleus

Vacuole

Lower cells of palisade

50μm

Fig. 1.5. Palisade cells of the privet leaf.

pathway for organic materials in solution (see fig. 1.6). The water-conducting elements of the system, the xylem, are even longer than the phloem elements. Their walls are heavily thickened with the complex phenotic substance, lignin, and lack protoplasmic contents in the mature state.

This brief summary of some highly developed cells and the tissues they form may serve to emphasize that while almost all cells have certain basic features in common, yet, if a plant is to be successful, some degree of specialization of cells and thus of tissue formation is a necessity. It is the net effect of all the cells together that enables a plant to survive and compete in any habitat.

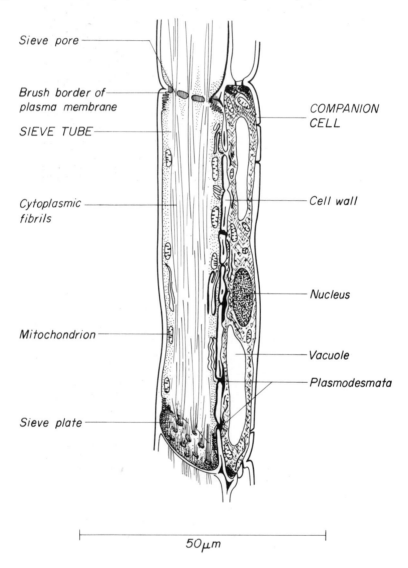

Sieve pore

Brush border of plasma membrane

SIEVE TUBE

Cytoplasmic fibrils

Mitochondrion

Sieve plate

COMPANION CELL

Cell wall

Nucleus

Vacuole

Plasmodesmata

50μm

Fig. 1.6. The sieve tube and companion cell of the phloem.

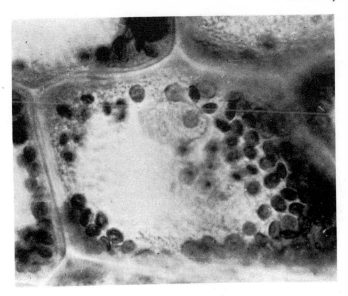

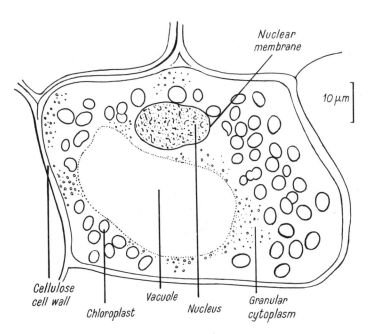

Fig. 1.7. A cell from the leaf of the Canadian pondweed (Elodea canadensis) showing the essential elements of a plant cell as seen with the light microscope.

General texts for further reading

ASHBY, M. (1961). *Introduction to Plant Ecology*. London, Macmillan.

BRADBURY, S. (1976). *The Optical Microscope in Biology*. Institute of Biology's Studies in Biology No. 59. London, Edward Arnold.

CLOWES, F. A. L. and JUNIPER, B. E. (1968). *Plant Cells*. Oxford, Blackwell Scientific.

ETHERINGTON, J. R. *Plant Physiological Ecology* (In preparation). Institute of Biology's Studies in Biology Series. London, Edward Arnold.

FOGG, G. E. (1963). *The Growth of Plants*. Harmondsworth, Penguin Books.

GRIMSTONE, A. V. (1976). *The Electron Microscope in Biology* (2nd edn.). Institute of Biology's Studies in Biology No. 9. London, Edward Arnold.

HURRY, S. W. (1965). *The Microstructure of Cells*. London, John Murray.

JAMES, W. O. (1973). *An Introduction to Plant Physiology* (7th edn.). Oxford, Oxford University Press.

RICHARDSON, J. A. (1964). *Physics in Botany*. London, Pitman.

SKENE, M. (1955). *The Biology of Flowering Plants*. London, Sidgwick and Jackson.

STREET, H. E. (1963). *Plant Metabolism*. Oxford, Pergamon.

STREET, H. E., and ÖPIK, H. (1976). *The Physiology of Flowering Plants* (2nd edn.). London, Edward Arnold.

'The Living Cell'. *Scientific American*, **205** No. 3, Sept. 1961.

2

Water Relations

2.1 Water and plant distribution

Water is such a familiar substance that we are liable to forget some of its important properties, without which life as we know it might well be impossible. Of these properties perhaps the most important relate to the maintenance of environmental stability. The high specific heat of water means that it takes considerable energy to warm it up; equally it is slow to cool and so water in lakes, streams, the atmosphere and soil, apart from that in the plant itself, has the ability to buffer sudden temperature changes. The lower density of water as it approaches freezing point is another important physical property as this results in the deeper water remaining warmer and usually unfrozen. In this way primitive and present day forms of aquatic life have been preserved from freezing. On the other hand this expansion in freezing poses a problem for terrestrial plants as the cell will be ruptured by the formation of ice crystals. Many plants possess quite a strong solution of dissolved materials in their vacuoles which help to minimize the effects of very low temperatures by depressing the freezing point of the vacuolar contents.

The exacting requirement of the plant for water is widely recognized by farmer and horticulturalist as well as physiologist. In natural plant communities there are often clearly defined groups of plants, a particular group being adapted to a particular level of water availability. In temperate zones the class of plants that are most usual are referred to as *mesophytes* and these normally require water in the soil at most times of the year, while under conditions of more or less continuous water scarcity are found a group of specially adapted plants referred to as *xerophytes*. In both these groups the individual plant's water relations are of great importance. For instance, under natural conditions in a normal mesophyte community there are often zones of vegetation in which competition between one species and another must be very subtly controlled, and the ability of one species to survive is often related to water availability (though other factors, such as the mineral content and acidity of the soil, may also be important). In the plant communities of bogs, fens and by streams and ditches, there is often a clear zonation, one species being dominant over a small zone of particular soil water content, only to be replaced by another at a slightly different soil saturation level.

A typical fenland community in the Itchen Valley in Hampshire is shown in fig. 2.1. Here water availability rather than mineral content or soil acidity seems to be the most important factor controlling competition between the three strong-growing herbaceous species: nettle (*Urtica dioica*), hairy willow herb (*Epilobium hirsutum*) and reed grass (*Glyceria maxima*). Nettle is usually dominant

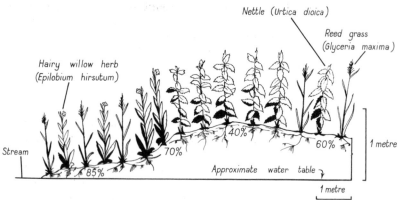

Fig. 2.1. An area of fenland in the Itchen Valley. Percentages indicate water content of the topsoil. (Class result.)

at 40 per cent water content; reed grass at above 85 per cent (though it has a wide range of toleration); in between these figures there is competition between all three species, hairy willow herb probably succeeding best at about 70 per cent saturation. A similar example of the importance of water availability (but often coupled with other factors, such as soil acidity) occurs in bog communities. An example of part of a valley bog at Pendle Hill in Lancashire in an area of high rainfall and on a base-deficient soil is shown in fig. 2.2. Here the bog-moss (*Sphagnum cuspidatum*) occupies the wettest niche and is usually found growing in open water. As humus accumulates, tussocks are gradually formed above the water level; these become colonized by other *Sphagnum* species, such as *S. plumulosum*, which is able to tolerate much more drying out. The driest places near the tops of the tussocks are colonized by *S. papillosum* and *S. palustre*, as well as by rushes (*Juncus squarrosus* and *J. conglomeratus*) and also by sedges and grasses (e.g. *Festuca ovina*).

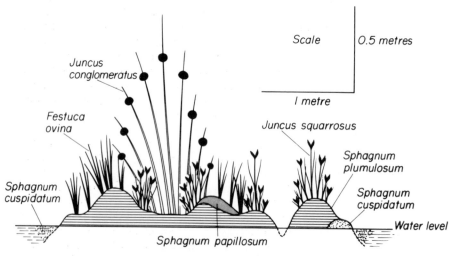

Fig. 2.2 A tussock in a valley bog, Pendle Hill, Lancashire.

Clearly then, the different plants are differently adapted with respect to their water requirements, and water availability is an important feature of the environment. It is important to know how water is utilized within the plant, as those species which can utilize water more efficiently will often have a competitive advantage.

2.2 The need for water

One of the most important requirements for water within the plant is to keep the cell protoplasm in its correct physical state. If the protoplasm is to be properly organized and its enzymes, microscopic and sub-microscopic organelles correctly functioning, these minute, colloidal particles must be able to circulate freely in a watery matrix. Some substances such as proteins will be inactived or *denatured* unless properly *hydrated* by water molecules. Dissolved mineral materials must be available at areas of synthesis, and dissolved gases must be able to diffuse into and out of the cell; all this can happen only if the protoplasm is fully hydrated. In addition, water is also important in maintaining cell turgidity by the inflation of the water-containing vacuole. This is particularly important in many annual and perennial non-woody mesophytes, which have little cell lignification and are largely dependent on the turgidity of their cortical parenchyma for support. Lack of water in such a mesophyte is usually immediately obvious as wilting occurs. The large quantity of water in the plant means that it is often subject to *frost damage*. Freezing causes the water in the vacuole to swell and the cell may rupture. Swelling of the water-containing elements of the cell as freezing approaches will cause a disruption of the metabolic activities of the cell but more serious and probably irrevocable damage will result if ice crystals pierce the organelle and cell membranes. If this happens the plant generally becomes a bright, dark green due to leakage of cell contents into the air spaces of the leaf. Wilting and ultimate browning of the leaf follow; the characteristic signs of frost damage.

Another important need for water is to make up for loss caused by transpiration from the leaves and to a lesser extent from the stem. This loss is often considerable, and in large mesophytic trees is reported to be as much as a thousand dm^3 on a hot day, while even in a semi-xerophytic palm tree, it may be four hundred dm^3 a day. Transpiration has a few positive uses. First, it will cause a lowering of temperature in the leaf due to the absorption of the latent heat of vaporization of water. Up to 80 per cent of the incident radiation is used in the evaporation of water and the resulting cooling of the leaf may prevent lethal temperatures being reached. By using small thermocouples attached to the leaf it has been possible to compare the temperature of wilted, non-transpiring leaves with those in an active state of transpiration. In some cases the cooling is only of a few degrees but in extreme cases such as the Californian monkey flower (*Mimulus cardinalis*), where the surrounding, ambient temperature may be as high as 60°C the temperature of the leaf may be kept at about 40°C. In this instance, as the temperature increases the stomata open more widely and the extra transpiration is effective in keeping the leaf cool. The flow of water from root to stem which is necessitated by transpiration is a more important effect and is referred to as the transpiration stream. This is the chief means by which dissolved minerals salts are transported upwards in the xylem vessels or tracheids from root to leaf.

Finally, least in magnitude but of great importance, water is required as a primary reactant in the process of photosynthesis in which the hydrogen made available is combined with carbon dioxide to synthesize carbohydrates. A number of important reactions also involve the addition and removal of water, for instance, the *hydrolysis* of starch to form sugar or the *condensation* processes by which sugar molecules join together to form starch.

2.3   Nature of the root system

If plants are to take up the water that they require they must, of course, have an efficient root system, well adapted to the ecological niche that the plant normally occupies. One of the most variable features in the structure of vascular plants is the root; even in one area there is a great deal of variation from one plant species to another, and also to some extent among plants of the same species. This is shown in the vertical transect through an area of chalk heathland (see fig. 2.3). Some plants, such as salad burnet (*Poterium sanguisorba*), have a distinct tap-root going down several inches into the soil. Others, such as ling (*Calluna vulgaris*) and the grasses have a fibrous rooting system, making the most of the water contained in the rich humus-containing layer nearer the surface.

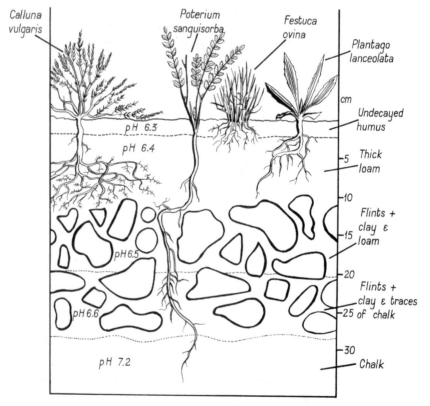

Fig. 2.3. Root systems of chalk heathland plants.

The actual area of water absorption is quite near the apex of the young growing root. There is very little cutinization or thickening near the apex, and water is able to diffuse easily into the internal spaces between the cortical cells. Many of the cells in this area are capable of taking up water, and so the total surface for water absorption is considerable. However, most water is taken up through the root hairs, whose other main function is anchorage. The main hairs develop in the outermost layers of cells, usually about 2 or 3 mm behind the root apex. Root-hair formation starts by a protrusion beginning to break through a weak point in the calcium pectate layer (see page 141) of the cell surface (fig. 2.4). The root hair elongates gradually, cellulose being continuously added at

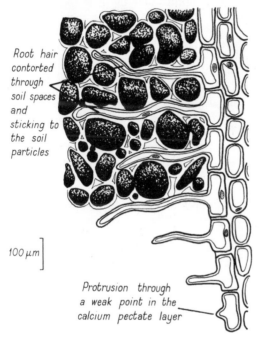

Root hair contorted through soil spaces and sticking to the soil particles

100 μm

Protrusion through a weak point in the calcium pectate layer

Fig. 2.4. Root hairs and their development.

the tip and the harder calcium pectate at the sides. Eventually the mature root hair is formed as a much convoluted thread, 3 or 4 mm long, winding through the spaces between the soil particles and anchoring the root firmly in the soil. As the whole root grows in length and thickness, there is usually a considerable friction between the root hairs and the soil, so that the life of a root hair is not very prolonged and their zone often ends after 5 or 6 mm. Plants grown in soil-less culture for instance, on moist filter-paper produce roots with much larger root-hair systems because of the lack of abrasion. On the other hand if grown in water-culture, root hairs may be entirely lacking.

2.4. Means of uptake of water and dissolved substances

There are essentially four ways in which materials, including water, may enter or leave a plant cell. In all cases movement will involve the expenditure or loss of energy. First there are a number of essentially physical processes; perhaps the most important of these is the process of *diffusion*, the energy for it deriving from the random spontaneous movement of ions and molecules. This process is important in the plant cell where there is a concentration difference between the substances inside and outside the cell. There will be a tendency for ions and molecules to diffuse down a concentration gradient until an equilibrium position is reached. Much of the uptake of water into the plant root takes place through the physical process of *osmosis* which may be regarded as a special instance of diffusion. Osmosis is the process by which water or solvent passes through a semi-permeable membrane to dilute a stronger solution. In principle the movement will continue until the concentration of both solvent (water) and solute (dissolved particles) is the same either side of the membrane, but in practice this is seldom so as pressures within the cell due to the elastic cell wall may prevent continued uptake. If internal pressures can prevent uptake then it is also possible that external pressure differences could force water into or out of the cell. These *hydrostatic pressure differences* are important in allowing, in part at least, for the *mass flow* of materials both from cell to cell and through the conducting elements of the whole plant. There are also a number of instances where the uptake or loss of materials may need the expenditure of respiratory energy; such processes are described as *active* and their links with respiration can be by the use of inhibitors of that process (see p. 79). The setting-up of diffusion gradients, osmotic gradients and hydrostatic pressure differences depend on both *external energy supplies*, such as that of the sun in causing evaporation at the leaf surface, and also to some extent on *internal energy supplies*, which may be directly linked to the metabolism of the cell. For instance the production of soluble, osmotically active glucose from insoluble starch will be likely to increase the osmotic uptake of water into the cell.

Osmosis is the chief means by which water is taken up into the root, and it is identical in principle to the physical demonstration of osmosis using the osmometer, in which the semi-permeable membrane is represented by a piece of bladder and the strong internal solution by potassium nitrate, copper sulphate or sugar solution. The pressure set up in the osmometer tube, or the force that has to be exerted to prevent osmosis taking place, is called the osmotic pressure (see fig. 2.5).

In the cells of the root tip and in living plant cells generally the cell wall is a fully permeable membrane, the cytoplasm semi-permeable and the vacuole contains a fair concentration of dissolved material. That the cytoplasm is the membrane can be shown by placing tissues such as the staminal hairs of *Tradescantia virginiana* (which contain a deep red-purple anthocyanin pigment in their cell vacuoles) in a strong external solution (e.g. molar potassium nitrate) so as to cause *plasmolysis* (see fig. 2.6). As the vacuole shrinks due to water passing out into the stronger external solution the cytoplasm comes away from the cell wall, but the cell wall itself is not particularly distorted, indicating that it is freely permeable.

Within the plant cell there are two factors determining whether a plant takes up or loses water. First there are the dissolved substances in the vacuole which determine the osmotic pressure or, more correctly, the *osmotic potential* of the cell sap. Second there is the elastic cell wall which exerts a force, the *wall pressure*, on the cell contents. In terms of the pressures *within the cell* and thus the potential for

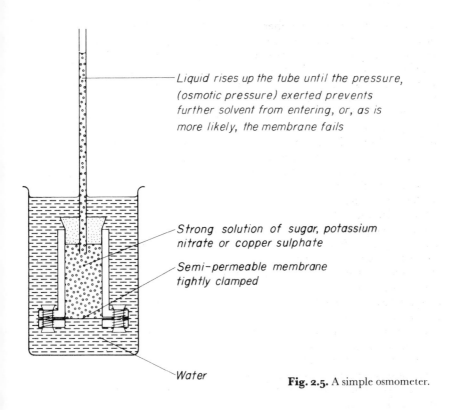

Liquid rises up the tube until the pressure, (osmotic pressure) exerted prevents further solvent from entering, or, as is more likely, the membrane fails

Strong solution of sugar, potassium nitrate or copper sulphate

Semi-permeable membrane tightly clamped

Water

Fig. 2.5. A simple osmometer.

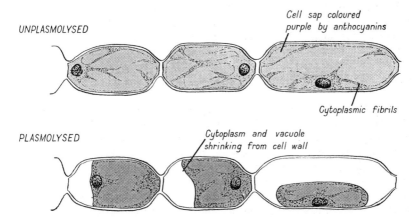

UNPLASMOLYSED

Cell sap coloured purple by anthocyanins

Cytoplasmic fibrils

PLASMOLYSED

Cytoplasm and vacuole shrinking from cell wall

Fig. 2.6. Plasmolysis of *Tradescantia* staminal hairs. A normal hair and the same after 20 mins in molar potassium nitrate.

water movement, the osmotic potential should be designated *negatively* (see fig. 2.7) as it will allow for water movement *into the cell*. The wall pressure on the other hand exerts a *positive pressure* on the cell contents which may even, under extreme conditions, contribute to an exudation from the cell.

A plant is spoken of as being fully turgid when its cells are fully distended and no more water is taken up. At this point wall pressure equals the osmotic potential of the cell sap. At full turgor the force distending the cell wall, or *turgor pressure*, is equal and opposite to the wall pressure, and therefore the same as the osmotic potential of the cell sap (fig. 2.7). If the cell is incompletely turgid more water may be taken up. This uptake of water depends on the difference between the osmotic potential and wall pressure. Formerly this was called the *suction pressure*, but this term has been replaced by the term *water diffusion potential* or simply *water potential*.

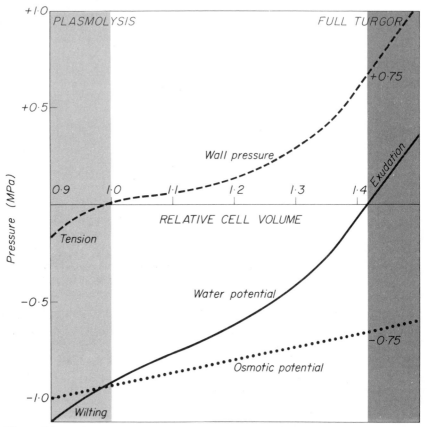

Fig. 2.7 Diagram (approximate pressures only) to illustrate changes in osmotic potential, wall pressure and water potential as conditions vary from plasmolysis to full turgor. (After Weatherley, P. E. (1952). Some theoretical considerations of cell water relations. *Annals of Botany*, N.S. XVI, 62).

The water potential approaches zero as the cell becomes more turgid and the wall pressure increases. During this take-up the cell sap becomes progressively diluted and consequently the osmotic potential becomes less negative. At full turgor the negative osmotic potential exactly balances the positive wall pressure and so the water potential is zero. Wilting and exudation are also expressed in fig. 2.7.

Measurements of the osmotic potential of the cell sap are usually taken at incipient plasmolysis, that is when plasmolysis is just visible in about half the cells of the tissue being examined (see Appendix, p. 202). At this point water potential is the same as the osmotic potential, the wall pressure being nil. Other methods for measurement of the water potential of the cell sap make use of changes in weight or size of tissues when placed in solutions of different strengths (see Appendix, p. 203). There will be no change in size or weight of a tissue or cell when it is placed in a solution which is isotonic, or of the same water potential as the tissue. Typical results are shown in fig. 2.8. For the corona of a daffodil flower

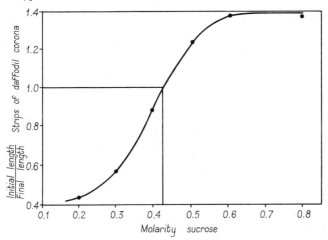

Fig. 2.8. Relationship between concentration of external solution and tissue size. Where the initial/final length of the strip equals unity, the molarity of the sucrose is equivalent to the water potential. In this case the molarity is 0.42. At 20°C. this represents a water potential of 1.2 MPa (see Appendix, p. 203). (Class result.)

the isotonic solution was 0.42M sucrose. For the accurate calculation of the water potential a knowledge of the temperature and the degree of ionization of the particular external solution at the particular temperature is necessary. Using unionized substances, such as sucrose at 0°C., the pressure exerted by molar solution would be expected to be 2.27 MPa*. Osmotic potentials at other temperatures can be calculated from the expression:

$$\text{O.P.} = \frac{273+t}{273} \times 2.27 \text{ where } t = \text{room temperature in } °C.$$

In practice experimental values (see p. 203) often differ considerably from those derived from this equation. The water potential of a 0.42M solution of sucrose is 1.2 MPa atmospheres at 20°C.

2.5 Permeability of the cytoplasm

The phenomenon of plasmolysis indicates the importance of the cytoplasm as the semi-permeable membrane; some idea of its physical nature can be obtained

* 1 MPa = 1.0×10^3 kPa ≃ 10 atmospheres.

from experiments relating the effects of external conditions to its permeability. One of the most important effects is that of temperature. As might be expected, knowing the cytoplasm to be a highly organized system of colloidal proteins, high temperature has the effect of destroying the semi-permeable nature of the cytoplasmic membrane. This is well shown by an experiment making use of the red anthocyanin pigment in the cell-sap of beetroot (see Appendix, p. 203). If similar-sized pieces of beetroot are immersed in water at a set temperature for one minute and then placed in distilled water for half an hour, the water is left coloured by the anthocyanins which have diffused out through the cytoplasm. Temperatures near boiling cause a deep red colour to be released, but at 60°C. the amount of red colour diffusing out falls off almost completely. This is the maximum temperature at which the cytoplasm can be maintained in its properly organized physical state, capable of acting as a semi-permeable membrane. This temperature is close to that which causes the coagulation of proteins and the breakdown of their colloidal state. Other more drastic treatments, for instance the use of a dehydrating agent such as alcohol, also cause the cytoplasm to become fully permeable. These experiments do not tell us what part of the cytoplasm is the actual semi-permeable membrane. Although experimental evidence is lacking, it could be one of three possibilities: first, the outer cytoplasmic membrane, the *plasma membrane*; secondly, the whole cytoplasm itself, and finally, the *tonoplast* or inner cytoplasmic membrane around the vacuole. Membranes are discussed further on p. 147.

In addition to the various functions of the cytoplasm in metabolism, that of acting as a semi-permeable membrane, and thus allowing the physical process of osmosis to take place, must rank as vitally important to the plant.

2.6　Water loss

Environmental factors affecting the rate of transpiration

There are two main techniques for investigating the rate of transpiration. The simpler methods involve weighing a whole plant, cut shoot or leaf. This will give the water loss in a given time. Alternatively, a potometer (fig. 2.9) may be used. This is an apparatus designed to estimate the rate of water uptake of an intact plant, or as is more usual, of a cut shoot, by observing the flow of water along a capillary tube to which the shoot is attached. Careful use of the potometer indicates the importance of the environmental conditions in determining the rate of transpiration or water loss. At the same time the rate of evaporation can be measured by use of the atmometer, a similar apparatus, but using a porous pot as the evaporating surface. It would be expected that the water loss from the plant would be on a parallel with the rate of evaporation and, provided that the other environmental conditions are constant, this is usually so. The rate of transpiration varies very much in the same way as the external physical conditions of temperature, atmospheric humidity and wind velocity affect the rate of evaporation from the purely physical system. If the effect of light is examined it is found that the rates of evaporation and transpiration are quite differently affected. Light itself has little effect on the rate of evaporation, but in most mesophytes the rate of transpiration is much less in the dark than the light. This is well illustrated by the diurnal fluctuations in transpiration rate shown by

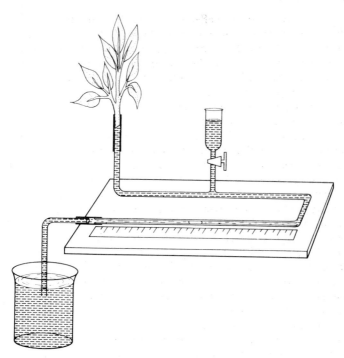

Fig. 2.9. The potometer. An apparatus used to measure the rate of water uptake by a transpiring shoot.

hairy willow herb (*Epilobium hirsutum*) (see fig. 2.10 and also Appendix, p. 204, for a full description of the methods used). It is, of course, the effect of light in causing the opening and closing of the stomata that is the basis of this diurnal fluctuation.

Structural features of the plant which may affect the transpiration rate

Before looking at the mechanisms which control the opening and closing of the stomata it is worthwhile discussing the various structural and anatomical features of the plant which may also affect the amount of water lost. The whole growth form of the plant is of course important; larger trees and shrubs are for the most part bound to have greater water loss than herbaceous and cushion plants. The total surface area of the leaves and the size of the individual leaf are also important. Plants living in humid areas lose water less readily by transpiration and are usually found with relatively large and expanded leaves; in some tropical rainforest deciduous trees the enormous pinnate leaves may be a metre long. On the other hand, plants of dry heaths and high alpine regions may have minute leaves.

More common adaptations which prevent excessive water loss concern the surface of the leaf itself. Cutin, a derivative of fatty acids, is almost impermeable to water vapour and, provided that the cuticular layer is sufficiently thick, loss of water through the cuticle is not thought to be particularly important. In most plants, though, the amount of cutin covering the lower surface of the leaf is considerably less than is found in the upper, and it is possible that here cuticular transpiration may be more important. Continuous accurate weighing of a single leaf over a period of time has indicated the magnitude of cuticular water loss.

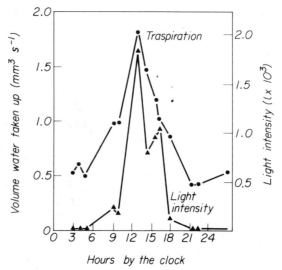

Fig. 2.10. Diurnal fluctuations in transpiration rate shown plotted with light intensity. Data from estimation of the transpiration rate using the potometer and a cut piece of hairy willow herb (*Epilobium hirsutum*) of 390 cm² leaf surface. Light intensity is measured by a photo-voltaic cell; temperature and evaporation rate can be measured at the same time. (See Appendix, p. 204, for a full description of the methods used.)

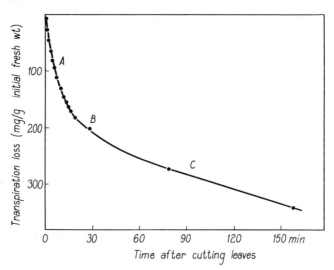

Fig. 2.11. Transpiration decline curve of ragwort (*Senecio jacobaea*). Leaves collected at 10.55 hrs (After Willis, A. J., and Jefferies, R. L. (1963). 'Investigations of the water relations of sand-dune plants.' *B.E.S. Symposium.*)

The transpiration rate may show a characteristic decline curve (fig. 2.11). This can often be divided into three phases; first (A) being the rate of transpiration with the stomata open; at (B) the stomata close, and the last part of the curve (C) gives the rate of water loss through the cuticle. In this case the cuticular loss is about a fifth of the total transpiration.

Electron-microscope photographs of carbon replicas of leaf surfaces show a great variety of surface; fig. 2.12 shows the layered protective system of waxy cutin covering found on spruce needles and *Kleina articulata*.

Fig. 2.12 The surface of (a) a spruce needle and (b) a leaf from *Kleinia articulata* (×28000). Electron micrographs of carbon replicas of surface. (Courtesy of Dr. B. E. Juniper.)

Hairy leaves are also frequently found in flowering plants, and the hairs are generally thought to help keep the leaf cool and also to prevent rapid wind currents from passing close to the surface and thus removing water vapour from transpiring areas. Although hairiness is widespread in mesophytes, it is particularly frequent in plants growing at high altitudes, where high winds and high temperatures are frequently encountered. For instance, alpine mouse-ear chickweed (*Cerastium alpinum*), a local alpine plant found in the north of England and Scotland, is densely covered with long white hairs. In the Alps, typical high-altitude cushion plants, such as *Androsace imbricata*, are covered with a growth of short hairs, while *Eritrichium nanum* and edelweiss (*Leontopodium alpinum*) are densely pubescent.

As most mesophtyes have their stomata open in the light, allowing free gas transfer for photosynthesis and respiration, the stomatal distribution, shape and size are probably, second to the actual leaf area, the most important group of factors controlling water loss during the daytime. During the night the stomata are usually closed and cuticular transpiration may be relatively more important.

The distribution of stomata can be examined by making epidermal strips or by making a nail-varnish replica (see Appendix, p. 205). There is a surprisingly wide range of stomatal densities found in ordinary mesophytes. Some examples of stomatal frequencies are given in the table below:

| Species | | *Frequencies* per mm² | |
		Upper epidermis	*Lower epidermis*
Scots pine	*Pinus sylvestris*	120	120
Oak	*Quercus robur*	0	340
Sunflower	*Helianthus annuus*	120	175
Geranium	*Pelargonium zonale*	29	179

Table to show stomatal frequencies in some mesophytes. (Data from Meidner, H. and Mansfield, T. A., 1968. *Physiology of Stomata*. London, McGraw Hill.)

These stomatal densities are considerably higher than those found in many xerophytes though many succulents have quite high stomatal densities.

The size and shape of the stomata are important anatomical features influencing the rate of transpiration. In most mesophytes the stomata are as shown in fig. 2.13, that is, they are nearly flush with the surface of the leaf; even

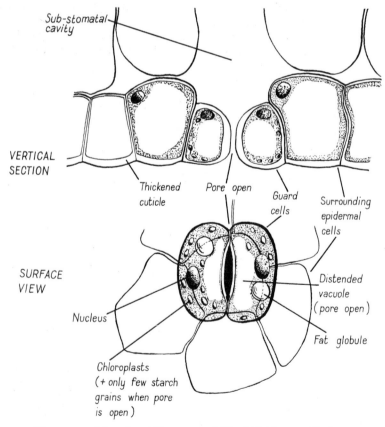

Fig. 2.13. The stoma of the privet (*Ligustrum ovalifolium*) highly magnified.

so, the actual area of opening and closing, between the guard cells, is very slightly sunk in many cases. (Extreme examples of the sunken stomata found in many xerophytes are discussed on page 40.) This small pocket of relatively still air may become highly saturated with water vapour, and this may result in an effective lowering in the rate of passage of water molecules from the saturated air spaces of the mesophyll out into the atmosphere. The size of the open pore itself is also important in determining the amount of water lost. It is subject to a good deal of variation from plant to plant, but typical dimensions, as in the garden rhododendron, are 0.006 mm long and 0.001 mm wide. However, Brown and Escombe found that a surface pierced by many small pores provides less resistance to the diffusion of water vapour than a similar surface pierced by a smaller number of larger holes of equivalent area. It is important, then, in comparing the stomatal organization of various plants, to consider the size of the pores in relation to their numbers and distribution.

The opening and closing of the stomata

A simple apparatus, called the porometer, can be used to measure the mass flow of gas through the leaf under different conditions. Early porometers such as that devised by Francis Darwin and Pertz in 1911 (see fig. 2.14) consisted of a cup sealed to the leaf surface (usually the lower surface). The time taken for the water in the vertical tube to fall gave a measurement of the rate of flow of gas through the leaf and indicated the resistance the leaf offered. A simple and efficient

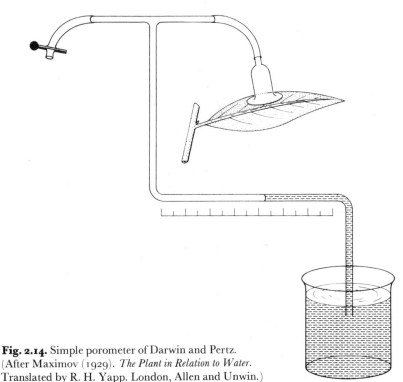

Fig. 2.14. Simple porometer of Darwin and Pertz.
(After Maximov (1929). *The Plant in Relation to Water.*
Translated by R. H. Yapp. London, Allen and Unwin.)

modern version is discussed in the Appendix on page 205. Using the porometer, the results illustrated in fig. 2.15 were obtained. Here the surface of the leaf offers much less resistance to the loss of gas during the daytime than at night, which agrees well with the visual evidence that the stomata are open in the light and closed in the dark.

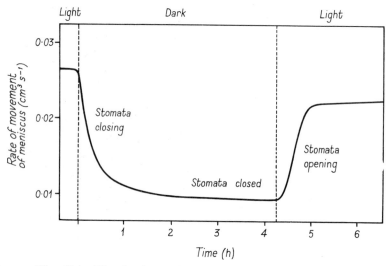

Fig. 2.15. The effect of illumination on the movement of the stomata of *Pelargonium*. (After Noel (1959). Some New Techniques in Plant Physiology, *School Science Review*, No. 142, p. 497.)

Stomata open when their guard cells are fully turgid and close when these cells lose full turgidity. This effect can be shown by placing an epidermal strip of a plant such as the garden iris in a strong solution of potassium nitrate, so as to cause plasmolysis of the guard cells and closing of the stomata. It is, however, usually only under conditions of drought and wilting that the stomata close on account of the overall water level within the plant.

Some mechanism must therefore exist which can bring about changes in the turgidity of the guard cells and the system must be dependant, directly or indirectly, on the influence of light. However, in laboratory investigations it has become clear that a number of other conditions may influence the opening and closing of the stomata.

A most important factor seems to be carbon dioxide; if the leaves are subjected to high carbon dioxide concentrations then closure occurs under most conditions but if the concentration is low then the stomata are usually open. This situation is probably linked to photosynthesis through the removal of the gas in the light. Similarly respiration during the night will allow for its accumulation. The most likely effect of this increase or decrease in carbon dioxide is to change the pH of the guard cells. Using indicators it has been found that guard cells at night may have a pH as low as 5 while during the daytime the value may increase beyond neutrality. Most enzymes (see p. 129) only operate over a narrow pH range and it is particularly interesting to find that some of the stomatal reactions could be modified through pH changes.

One possibility concerns the appearance and disappearance of starch in the guard cells. These, unlike most others in the green parts of the plant, accumulate starch in the dark and sugars in daylight (see fig. 2.16), the latter being formed as

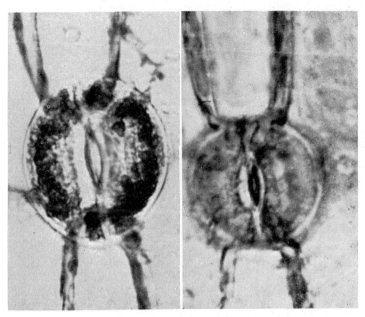

Fig. 2.16. Starch accumulation in the guard cells of *Iris*. *Left:* strong, accumulation after 24 hours in the dark; *right:* after 12 hours in the light. Stained in iodine. (Highly magnified.)

a product of guard cell photosynthesis. The enzyme *starch phosphorylase* (see p. 245) has been extracted from guard cells and is able to convert starch to glucose-1-phosphate, or the reverse according to the pH of the medium, the direction of the reaction being in accord with the observed pH values of the stomata; that is the synthesis of starch is favoured under acid conditions while that of sugar tends to take place at neutrality. Although glucose-1-phosphate is itself osmotically active, as it is broken down to glucose and phosphate ions then the osmotic potential of the cell sap will increase; on the other hand if starch accumulates the osmotic potential will fall as the substance is inactive osmotically. This change in osmotic potential will increase or decrease the amount of water taken up into the cell and consequently, through gain or loss of turgor the stomata will open or close.

The work of Mouravieff (1957) has shown that the starch-sugar reaction is influenced by blue light, presumably absorbed by the carotenoid pigments. Wavelengths around 459 nm cause stomatal opening and apparently assist in the conversion of starch into sugar, though it should be pointed out that some monocotyledonous plants such as onion, which do not form starch in their guard cells at all, are also influenced by blue light causing their stomata to open. Further recent work by Fujino (1967) has thrown yet a different slant on the situation as, by a simple staining technique using sodium cobaltinitrate (see

Appendix p. 206), there is now plenty of evidence to show that guard cells accumulate potassium to quite high concentrations. It could be that the sugar made available under light conditions is used, in part, for the active uptake of potassium from the surrounding cells. This would increase the osmotic potential of the cell sap and water would then be taken up.

There are many other complexities; why for instance do the stomata possess an innate rhythm of opening and closing so that if they are kept in continuous daylight the rhythm may persist for several days? In spite of these and many other enigmatic points it is still possible to put some of the situations described above in a logical sequence which may at least summarize the situation operating in most mesophytic plants:

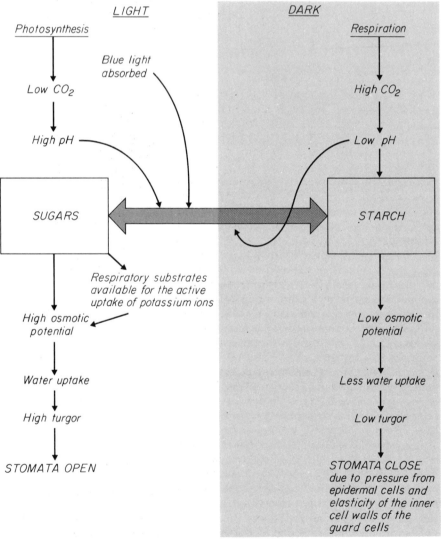

Summary of the conditions influencing the opening and closure of the stomata in mesophytes.

Water loss by guttation

When a plant is growing under conditions of high soil-water content and high atmospheric humidity small droplets of water are often seen at more or less regular intervals around the edge of the leaf. Such conditions are often noticed in a greenhouse after copious watering the previous evening followed by a relatively cool night. This exudation is called *guttation* and normally occurs through specialized structures called *hydathodes* which are found at the edges of the leaves close to the endings of the xylem vessels. This type of water loss, unlike transpiration, is to a large extent under metabolic control, and the process bears in this respect a similarity to the phenomenon referred to as root pressure, in which considerable water pressures are developed in the roots. How important it is to the plant it is difficult to say, but it seems likely that it allows for a continued flow of water and dissolved materials through the plant even when transpiration is low.

2.7   Transport of water

A normally turgid leaf contains water in its vacuoles, cytoplasm, cell walls and inter-cellular spaces. Transpiration results in the removal of water, primarily from the last two of these areas; these are referred to as the *outer-space* or *apoplast* (see fig. 2.19). The lowering of *hydrostatic pressure* caused by evaporation could result in water movement either from the vacuoles of leaf cells or from the vascular system via further cell walls and inter-cellular spaces. Weatherley has shown that relatively little water comes from the cell vacuoles and most of the movement is through the apoplast system. Movement of water into and out of the cells does still take place as alterations in their osmotic potentials occur; these can be brought about by metabolic processes such as photosynthesis. It is thought that such movement is small in magnitude compared with that occurring through the apoplast system. Loss of water from the finest xylem elements is made good by a corresponding uptake in the root. The xylem acts as a long and continuous tube from root to leaf, though its vessels and tracheids are very narrow, being of the order of 0.005 to 0.05 mm in diameter. Tall columns of water have an extreme ability to stand up to tensions without breaking, and it is thought that the sun's energy, causing the evaporation from the leaf, is enough to account for most of the ascent of water in both herbaceous plants and all trees. That this transpiration pull or *shoot-tension* can exert a considerable force is well shown by the experiment in which a shoot of cedar

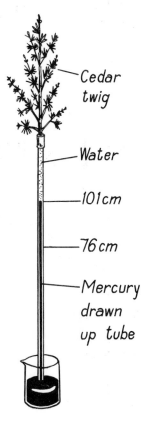

Labels in figure: Cedar twig — Water — 101cm — 76cm — Mercury drawn up tube

Fig. 2.17. Demonstration of shoot tension. (After Greulach (1957). 'The Rise of Water in Plants', *Plant Life*, New York: Scientific American.)

is attached to a long glass tube filled with water and dipped into mercury. With an atmospheric pressure of 76 cm of mercury, the mercury in the tube rose to over 101 cm, indicating a considerable force exerted by the plant (see fig. 2.17).

That most of the water is transported in the xylem rather than the phloem is also easily demonstrated by placing a cut shoot of a rapidly transpiring plant, such as the balsam or Busy-Lizzie (*Impatiens sultani*), in dilute aqueous eosin or methylene-blue dye. After a few hours sections cut high up the shoot will show that most of the dye is clearly in the xylem regions. This can be confirmed by another simple experiment, in which the cut end of a leafy shoot of a woody plant is prepared so that the xylem and phloem can be covered separately with wax.

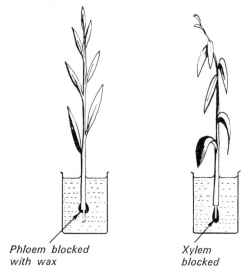

Phloem blocked   Xylem
with wax   blocked

Fig. 2.18. The path of movement of water in the stem. The effect of covering the xylem and phloem with wax.

With the xylem blocked, upwards transport of water is prevented and the plant wilts, but a shoot with the phloem blocked is unaffected (see fig. 2.18).

The lowering in pressure at the bottom of the finest xylem vessels will allow water to pass into them from surrounding areas. Again, there are two possibilities; water could be drawn from the apoplast of the root system or from the vacuoles. The work of Weatherley has shown that in this area water movement is subject to conditions affecting the cytoplasm (e.g., temperature) and so it seems likely that at least part of the movement involves transport of water *through* living cells. The semi-continuous cytoplasmic system is referred to collectively as the *symplast*. The root endodermis, which in dicotyledonous plants has a lignified and suberized *Casparian strip* and in monocotyledons other types of wall thickening, is one site where water must move almost exclusively through the symplast (see fig. 2.19). Much of the movement through the rest of the root could, however, still be through the apoplast. Apart from the endodermis, it is therefore possible to visualize a continuous apoplast from soil water surrounding the cells of the root apex, right through the plant to the surfaces of the leaf cells. While the bulk of this water movement is concerned in making good the transpiration loss, certain amounts of dissolved minerals, taken up from the soil by active, energy requiring processes, are transported in the main transpiration

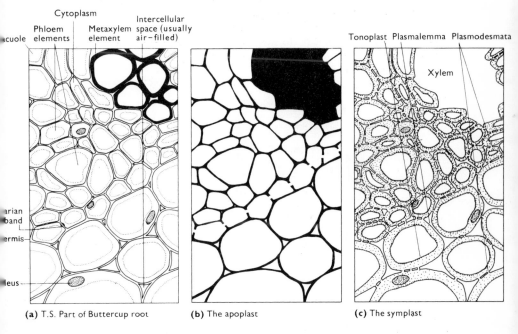

(a) T.S. Part of Buttercup root   (b) The apoplast   (c) The symplast

Fig. 2.19. TS of part of a buttercup root (**a**), illustrating the apoplast (**b**), and symplast (**c**). (After Baron W. M. M., *Water and Plant Life* (1967), London, Heinemann, as modified by Sutcliffe, J. F. and Baker, D., *Plants and Mineral Salts* (1974). Studies in Biology No. 48, London, Edward Arnold.)

stream. In this way they are carried near to areas such as the stem apex where active metabolism is occurring.

In summarizing it should be emphasized that water will only move down a *gradient of water potential*. This may be due to all kinds of external and internal factors; external factors such as temperature through causing evaporation at the leaves may reduce the pressure at that site; internal osmotic conditions may affect the gradient either way.

2.8   Transport of dissolved organic materials

Although the movement of organic materials may seem unrelated to water relations and water movement, there is considerable interrelationship between the two aspects of translocation. Although some organic materials may be transported in the xylem most are moved through the symplast system, together with some water and dissolved mineral ions. In such areas as the leaf mesophyll and root cortex much movement will occur from cell to cell through the plasmodesmata (see p. 2) but movement over greater distances will usually involve the vascular system of the phloem. Evidence that the phloem is the main pathway for the conduction of such substances comes from ringing experiments and the use of radioactive tracers. Woody stems have their phloem under their bark, just outside the woody tissues of the xylem and separated from them only by the thin layer of cambium. Cutting out a ring of phloem does not kill the tree immediately as water can still pass up the xylem from root to leaves (see fig. 2.20) but eventually the tree will die owing to starvation of the roots.

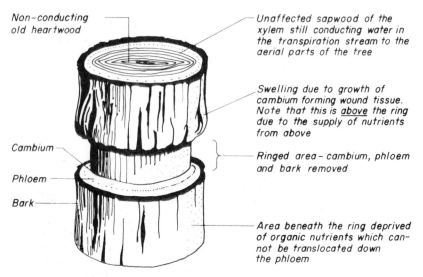

Non-conducting old heartwood

Unaffected sapwood of the xylem still conducting water in the transpiration stream to the aerial parts of the tree

Swelling due to growth of cambium forming wound tissue. Note that this is _above_ the ring due to the supply of nutrients from above

Cambium

Ringed area - cambium, phloem and bark removed

Phloem

Bark

Area beneath the ring deprived of organic nutrients which cannot be translocated down the phloem

Fig. 2.20. Part of a tree trunk that has been ringed some months previously.

If organic materials are stored in the roots or in underground tubers then there may be times when organic materials are translocated upwards to growing parts. Equally, leaves that are in a state of active photosynthesis may export organic substances to young, immature leaves and flower buds near the stem apex; such movement is also in the phloem. Although sugars are perhaps the commonest substances translocated, a wide variety of organic molecules are also moved. Some of the most important of these are the *plant hormones* (see chapter 7) which exercise such a vital role in the regulation of growth and development. Whether the phloem itself acts in a *regulatory capacity* is difficult to say but the fact that the phloem cells are alive suggests that they may be able to exercise control over the rates, at least, at which such vital materials are moved about the plant. It is difficult to see how substances can be moved through the long, living, cytoplasm containing cells of the phloem (see fig. 1.6) at such high velocities as experiments with radioactive tracers have shown: some sugars may move at the rate of 100 cm an hour. Phloem cells are notoriously easily upset; poisons and changes in oxygen tension and temperature cause immediate and often profound effects on their ability to translocate substances, and so it seems likely that the living system of the phloem must in some way be responsible for the movement. Some energy requiring process must actively pump or direct the materials along the sieve tubes. Much research has been carried out to try to understand the mechanism of movement but, as so often happens in physiological systems, the situation is made particularly complex through the possibility that a number of different systems may operate; perhaps separately in some instances, perhaps together at other times.

For instance, Münch (1930) has suggested that a pressure flow system operates. This is illustrated in the model shown in fig. 2.21. The model consists of

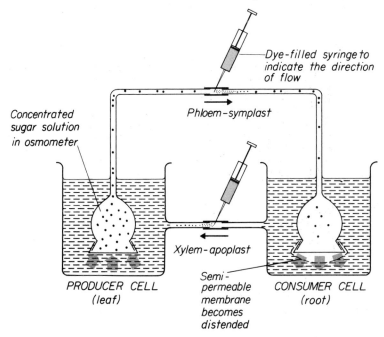

Fig. 2.21. Münch's model to illustrate how a pressure-flow system could operate in the xylem and phloem.

two osmometers connected as shown. One represents a *producer* cell and contains a high concentration of sugars, the products of photosynthesis. This is connected by a tube to the second osmometer which represents a *consumer* cell. The connecting system represents the symplast or phloem. Material will pass down the concentration gradient from producer to consumer. The apoplast is represented by the water surrounding both 'cells'. Water will enter the producer cell by osmosis causing a pressure to be set up inside that cell which will cause a mass flow towards the consumer cell. Equally, the reduction of pressure in the surrounding apoplast will cause a mass flow along the 'xylem'. If such a model is set up it will work well until the concentration of materials in both osmometers becomes the same, a situation which should not occur in a living plant if photosynthesis continues in the leaf and if the consumer cell is either breaking down the sugars through respiration or converting them to insoluble and osmotically inactive substances such as starch.

Münch's ideas undoubtedly have much to recommend them, not least in that they show the need to think of interacting systems rather than necessarily individual cells and much work has gone on to try to see whether his model is truly applicable to the living plant. For instance, Zimmerman has been able to carry out an elegant series of experiments using aphids. These insects derive their food by sucking the sap from plant leaves and stems. They are able to bore into the sieve tubes using their microscopic stylet mouthparts without much upsetting the living cells. The animal can be cut away from its mouthparts, leaving the stylet in place connected to the sieve tube. In this way it is possible to sample their contents. It was found that exudation continued from the stylet, which supported Münch's hypothesis that the phloem contents should be under a positive pressure. Zimmerman analysed the contents of the sieve tubes and found that there is a progressive decline in the concentration of sugars from the photosynthesizing parts of the plant downwards. This would also support the idea of the producing and consuming cells. Other investigators have found that when the xylem is punctured there may be a sudden hiss of entering air; this supports the concept of the pressure within the xylem system being less than atmospheric.

One of the drawbacks to Münch's hypothesis is that experiments with radioactive tracers such as ^{32}P and ^{14}C (which can be distinguished fairly easily due both to their differing half-lives and also the differences in the strengths of their β-emission) have shown that substances containing these elements can move in *different directions*, even in the same sieve tube. Recently much careful observation of sieve tubes has shown that the phenomenon of protoplasmic streaming is more common than used to be thought; it has been suggested that this energy-requiring process could allow for the bi-directional movement. Work by Thaine (1969) and others has suggested that these visible streaming movements as well as the movement of organic materials in general may be due to undulations and contractions in strands of protein running through the sieve tubes and frequently through the interconnecting pores. These strands may be composed, in part at least, of endoplasmic reticulum sheathed in specialized protein fibres which are thought to be contractile in nature. The organic material that is being translocated could pass alongside or within the strand, may be forced along by some kind of peristaltic pump. In addition there is usually a brush-border of small micro-villi projecting inwards from the plasma membrane adjacent to the sieve plate; it is tempting to suggest that this is also involved in the active movement of dissolved substances.

Like a number of physiological processes, that of translocation in the phloem looks as though it could operate in a number of ways, maybe operating simultaneously, but there is no doubt that it is an energy-requiring process and one which requires a high level of cellular organization.

2.9   Halophytes and xerophytes

Not all plants live in conditions where water is readily available throughout a great part of the year; such plants are particularly useful in studying water relations, as they may have interesting adaptations of considerable survival value in their special environments. There are essentially two classes of plants which can tolerate lack of available water. In the first case there are the *xerophytes* which are adapted to survive conditions of actual drought and water scarcity. In the second case there are those plants which may be living in conditions where there is plenty of water, but this contains so much dissolved material that normal mesophytes are unable to survive as their cells become plasmolysed. Plants living in areas with a high concentration of sodium chloride such as salt marshes are called *halophytes*.

The halophytes

The most usual adaptation in halophytes is a physiological one. That is, their cells have a much higher osmotic potential than is found in mesophytes, and in this way the plants are able to obtain their water under most conditions. Fig. 2.22

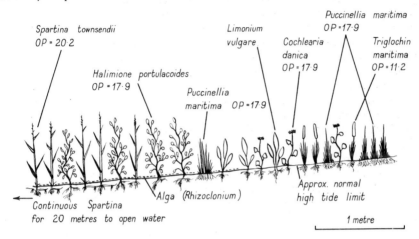

Fig. 2.22. Salt marsh plants and the osmotic potential of their cell sap, Keyhaven, Hampshire. (Osmotic potentials are given in atmospheres.)

shows a transect across part of a salt marsh near Keyhaven in Hampshire. Details of the osmotic potential of their leaves measured at incipient plasmolysis are also given.

Salt-marsh plants are adapted to their environment in several other ways. A common adaptation is that they often have a well-developed rhizome and rooting system which helps them to resist tidal action and bind the mud. Many of the grasses are specialized in this way; an outstanding example is the cord-grass, *Spartina townsendii*, which has colonized vast areas of mudflat around the British Isles. Many of the smaller plants frequently have succulent leaves; in some cases, such as the scarlet pimpernel, *Anagallis arvensis ssp. arvensis*, the succulence may be only an environmental modification which is lost when the plant is transferred to an area which does not contain salt in the soil. In other cases, such as the annual glassworts (*Salicornia* species), the plants are always succulent. Succulence is an important adaptation which helps the plants to survive in their unusual ecological niche. At some times of the day the plants may be bathed in sea-water of fairly low osmotic potential, but at low tide considerable drying out may occur, and this will result in the osmotic potential of the soil solution rising considerably. If the concentration of dissolved substances in the soil solution becomes higher than that in the plant, then 'physiological drought' will occur and the plant will be unable to take up more water, and water may even be withdrawn from the plant. Storage of water in succulent leaves may help the plant to tide over these difficult times of the day. Halophytes, then, are a group of plants which have several interesting adaptations that enable them to obtain the water that they require for life in what might appear to be a rather unfavourable environment.

The xerophytes

Xerophytes are plants which show a variety of adaptations enabling them to survive conditions of drought and water scarcity. They are found in a range of species from the true cacti of the *Cactaceae*, often plants of arid deserts, through the succulent species with swollen stems and leaves, to those plants which have no obvious external adaptations but which are nevertheless well adapted physiologically.

The xerophytes are best divided from the point of view of their various adaptations into two classes, the *drought evaders* and the *drought endurers*. The former class consists of those plants which really evade the issue and survive times of drought by existing as a seed or even as a spore. Much of the famous desert ephemeral vegetation is made up of annuals which come to flowering quickly in a wet period and then survive the ensuing dry one in the form of seeds. The Californian poppy (*Eschscholtzia californica*) is a well-known example of such a drought evader.

Truer xerophytic adaptations are found in those plants which are adapted to *endure* water loss. In several cases, such as the creosote bush (*Larrea divaricata*), a North American desert plant, the plant looks superficially much like any mesophyte, but has a high tolerance of desiccation, the protoplasm being able to retain its organization when the whole plant is in a highly dehydrated state. Many members of the Bryophyta and Pteridophyta have a rather unexpected tolerance to desiccation of this sort. For instance many mosses live in particularly dry areas; *Tortula muralis* is a common moss growing on walls and on rocks in the British Isles, and although it may appear to be completely dried out, revives quickly after a shower of rain. There are also several examples of Pteridophytes that are similarly able to withstand desiccation; the American resurrection plant (*Selaginella lepidophylla*) is often sold as a curiosity, as it exists in a rolled-up state when dry and rapidly unrolls when moistened. The ability of plants to stand up to protoplasmic desiccation is very considerable and is probably also found in a great many xerophytes with other adaptations.

Another means for the endurance of water loss is to have a large water storage system which enables the plant to survive the period of drought. This is found in many succulents which may have a high transpiration rate but which may still be able to survive in a desert region. In Britain many stonecrops (*Sedum* species) are in this class and are common plants of roofs, walls and rocks. Water is stored in large parenchymatous, frequently mucilage-containing cells of the leaf and stem. Endurance of water loss may also be possible if the plant has a sufficiently deep root system to enable it to obtain water throughout the period of drought. Such a system is particularly common in many Mediterranean trees and shrubs, such as the acacia and oleander.

The most interesting group of xerophytes is that group of drought endurers which is adapted to prevent water loss. There are a great many adaptations towards this end, but the simplest means of reducing water loss concerns reduction of the surface: volume ratio of the whole plant (and this at the same time usually results in succulence). Plants can normally be classified into stem and leaf succulents on this basis. The least well-adapted in the leaf-succulent series are possibly species of *Sedum* (see fig. 2.23). Here the leaves are swollen but

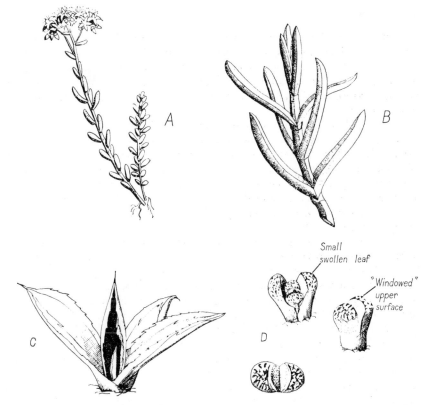

Small swollen leaf

"Windowed" upper surface

Fig. 2.23. A series of leaf succulents. (×½.) (A) The white stonecrop (*Sedum album*). (B) The Hottentot fig (*Carpobrotus edulis*). (C) The century plant (*Agave americana*) (×⅕) (D) A pebble plant (*Lithops helmuti*). An extreme form of leaf succulent. The chlorophyllous cells are deep down lining the leaf interior and the upper surface is nearly transparent. Many *Lithops* are nearly buried in pebbles but are still able to photosynthesize through this 'window' arrangement.

there is little lowering of the area from which transpiration may occur. The Hottentot fig (*Carpobrotus edulis*) is better adapted. In this the opposite leaves are closely packed down on one another. Other well-adapted species take the form of a rosette; the magnificent species of *Agave* that are common on the hillsides of the south of France are good examples of this class. The best-adapted of the leaf-succulent series is the genus *Lithops*. These look like pebbles and are found embedded in the soil of the Kalahari desert in South Africa. In *Lithops helmuti* the leaves are very swollen and sunk deeply in the soil; the upper surface of the leaf is transparent (see fig. 2.23) and thus allows for more efficient photosynthesis by the cells lining the inside edges of the leaves. For this reason they are often called 'window-plants'.

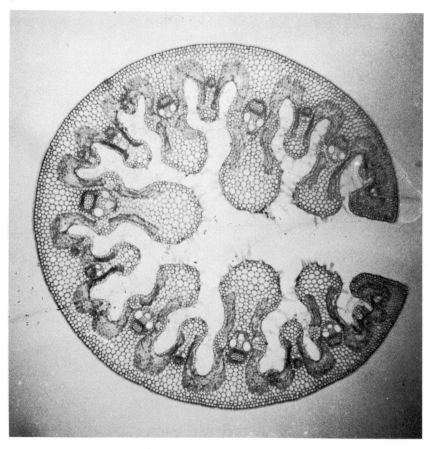

Fig. 2.24. T.S. leaf of Marram Grass (*Ammophila arenaria*) in rolled-up position reducing water loss. (Courtesy of A. J. M. Ventris.)

In the stem-succulent series the simplest type of adaptation is seen in *Kleinia articulata*, the candle plant; this has well-developed, though slightly succulent leaves and a thick, succulent stem, the leaves being shed during periods of drought though if treated with long-day conditions (see p. 189) the leaves will still be shed even if the plant is well supplied with water (see fig. 2.25). Leaf-shedding is a common feature of both mesophytes and xerophytes in times of extreme drought. The most famous example of stem succulents are in the cactus family itself; the best adapted species are found in the genera *Carnegiea* and *Cereus*. In these the stem is a simple, grooved upright structure with only occasional branches, true leaves are absent, but their place is taken by spines which line the edges of the stem. In some stem-succulents the adaptations have, in a sense, gone too far and the stems have a flattened and almost leaf-like appearance; *Opuntia ficus-indica*, which forms a large bushy plant with rounded stem segments up to 0.4 m long, and the smaller *Epiphyllum* species fall into this class. The competitive ability of some of the prickly pears (*Opuntia inermis* in particular) is well known. These plants colonized vast areas of Australia—some

60 millions acres—before they were brought under control by the larva of the moth *Cactoblastis cactorum*.

In addition to these gross adaptations which give the plant a small surface from which transpiration may take place, many xerophytes have other less

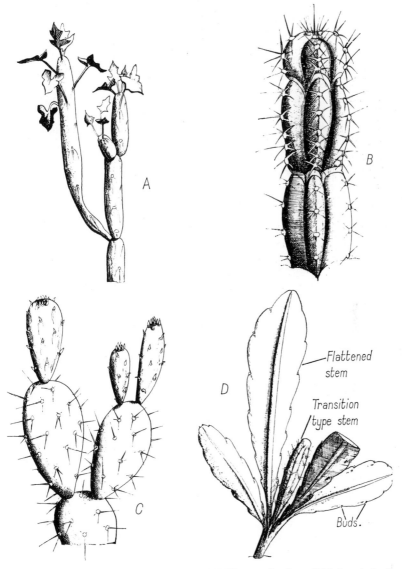

Fig. 2.25. A series of stem succulents. (×⅓.) (A) The candle plant (*Kleinia articulata*). The leaves are shed in times of drought. Water is stored in the swollen stem. (B) *Cereus peruvianus*. No true leaves are found, but their position is taken by spines. (C) Prickly pear (*Opuntia ficus-indica*). The oval, flattened stem segments are almost leaf-like in appearance. (D) Orchid cactus (*Epiphyllum crenatum*). An extreme example of a stem-succulent in which many of the stem segments are most leaf-like in appearance.

conspicuous adaptations which cut down water loss. A thick, waxy cuticle, as is found in *Kleinia*, is useful in reducing cuticular transpiration, while loss through the stomata is cut down primarily through their density being much less than is found in ordinary mesophytes (p. 24). For instance, *Opuntia ficus-indica* has about 100 mm^{-2}, while *Cereus chende* has only about 30 mm^{-2}. Loss through the stomata is also reduced by their arrangement in pits, sometimes surrounded by a system of hairs. This is well illustrated by the marram grass (*Ammophila arenaria*), which rolls its leaf up under dry conditions so that water molecules, diffusing out of the stomata, follow an indirect and lengthy route (see fig. 2.24).

Many xerophytes, particularly succulents allied to *Mesembryanthemum*, possess an interesting physiological condition in that they have an inverted stomatal rhythm, their stomata being closed during the day and open at night. This adaptation may be of particular importance as a method for water conservation, provided that sufficient carbon dioxide assimilation is possible during daylight, when photosynthesis is taking place (see also p. 65).

Xerophytes and halophytes have a series of adaptations which appear to be useful in their particular environments. In temperate-region plant communities water is still one of the most important environmental factors, and the distribution of many species may be shown to be closely correlated with water availability, though it is usually much more difficult to find what features mesophytes possess which enable them to compete successfully for water.

2.10   Water and the Ecosystem

No discussion of water relations and transpiration would be complete without mention of the effects of water loss on climate. Much of the incoming radiation from the sun is used in warming the Earth's atmosphere, causing water to evaporate and providing energy for wind and water currents; some of it is also reflected. In Britain of the total energy entering the atmosphere some 16 per cent may be absorbed by plants; of this energy only 1–5 per cent is used in photosynthesis and the rest is lost either by radiation or by being used in evaporating water. This is summarized in the table below (approximate average figures modified from Phillipson (1966)).

| | | Percentage of energy | |
	$MJ\ m^{-2}y^{-1}$	entering atmosphere	absorbed by plants
Energy entering the Earth's atmosphere	6.4×10^3	100	—
Radiant energy absorbed by plants in Britain	10.5×10^2	16	100
Energy loss from plants due to evaporation and radiation	$10–10.4 \times 10^2$	15.2–15.8	95–99
Energy used in photosynthesis	1–5	0.2–0.8	1–5

Apart from indicating the considerable amount of energy used mainly in the process of transpiration these figures also suggest the considerable influence that transpiration must have on the climate, first, as mentioned on p. 11 through buffering temperature changes and second through increasing the humidity of the atmosphere. The magnitude of this water loss is difficult to visualize. A single maize plant may lose 200 dm³ of water in a season, so for a 20 ha (50 acre) field planted with 50 000 plants to the hectare the loss of water in a growing season will be 2 × 10⁸ dm³. The effect of this vast loss on both soil water and atmosphere is vitally important to the whole local ecosystem. Broadly speaking, the less arid and more humid the conditions, the more vigorous will be the growth of plants. This has considerable relevance to the reclamation of desert and near desert areas where the establishment of some kind, indeed, almost any kind of vegetation is a necessary first step. This not only increases the humidity of the area and therefore makes it feasible for other economically more desirable plants to grow, but it also makes it more suitable for clouds to form and in this way rainfall may increase so that yet more plants will be able to grow. Unfortunately, the opposite is too often the case as removal of forests may lower humidities, reduce precipitation and produce a situation of incipient desert, often exacerbated by speeding up of run-off through the removal of the tree roots so that soil erosion ensues.

Further reading on Water Relations

BARON, W. M. M. (1967). *Water and Plant Life*. London, Heinemann.
MAXIMOV, N. A. (1929). *The Plant in Relation to Water*. London, Allen and Unwin.
MEIDNER, H. and MANSFIELD, T. A. *Physiology of Stomata*. London, McGraw-Hill.
RICHARDSON, M. (1975). *Translocation in Plants*, 2nd edn. Institute of Biology's Studies in Biology no. 10. London, Edward Arnold.
RUTTER, A. J. (1971). *Transpiration*. Oxford Biology Reader no. 24. Oxford, Oxford University Press.
RUTTER, A. J. and WHITEHEAD, F. H. (1963). *The Water Relations of Plants*. Oxford, Blackwell.
Society of Experimental Biology (1965). *The State and Movement of Water in Living Organisms*. S.E.B. Symposium no. 19. London, Cambridge University Press.
SUTCLIFFE, J. (1977). *Plants and Water* (2nd edn.). Institute of Biology's Studies in Biology no. 14. London, Edward Arnold.
WOODING, F. B. P. (1971). *Phloem*. Oxford Biology Reader no. 15. Oxford, Oxford University Press.

3

Photosynthesis

3.1 Introduction

Although water availability is one of the most obvious factors influencing plant distribution, light is probably just as important an ecological factor. Both light intensity and quality affect the rate at which plants can manufacture the complex organic materials that they require for their stores of energy. In woodland communities, for instance, plants are distributed according to light intensity. The dominant tree species are usually sun plants, thriving in high light intensities, but having seedlings which are usually capable, as in the oak, of tolerating shade conditions. The plants of herb and shrub layers differ markedly from one another in their light requirements, though many of them complete the main part of their life cycle in the spring before shading from the leaf canopy takes place. Respiration goes on all the time. When the light intensity is sufficient to allow photosynthesis to take place so that the rates of the two processes are the same, then the plant is said to be at its *compensation point*. At this point there is no net gain or loss of either carbon dioxide or oxygen. Shade plants can utilize low light intensities more efficiently than sun plants and so attain their compensation point earlier in the day (see fig. 3.1). Similarly, they are able to make up quicker

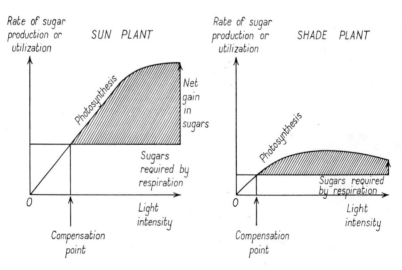

Fig. 3.1. The photosynthetic efficiency of sun and shade plants. (After Ashby: *Introduction to Plant Ecology.* London, Macmillan & Co.)

for the loss of carbohydrate that has occurred due to respiration during the dark. The time taken is called the *compensation period*. A plant that is slow in completing its compensation period will tend to be at a disadvantage compared with a faster neighbour, as new synthesis and growth may be delayed and the plant out-grown and out-competed. On the other hand, these shade plants which have a short compensation period are unable to utilize the high light intensities so efficiently (see fig. 3.2). Measurements of the compensation period of bryophytes growing on tree-trunks and branches have shown that light is an important factor in affecting the competition between species of rather similar habit growing on much the same substrate. Results of work on epiphytes growing on beech trees in Japan are given in the table on p. 44. A method used in determining the compensation period is given in the Appendix on p. 207.

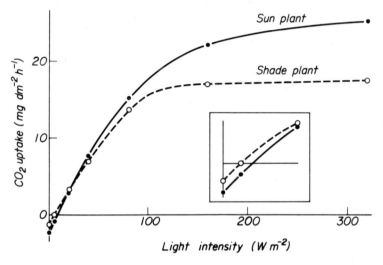

Fig. 3.2. The carbon dioxide exchange of leaves from Golden-rod (*Solidago virgaurea*) ecotypes adapted to exposed (sunny) or shaded habitats. The inset shows the lower part of the graph on a larger scale. (Adapted from Björkman and Holmgren (1963). Physiologia Plantarum Vol. 16.)

Although some plants may have slight anatomical differences when growing in the shade, for instance the number of palisade layers is often reduced, nevertheless, the mechanisms are largely unknown by which different plants are able to photosynthesize more effectively than others and obtain a competitive advantage.

Photosynthesis is not the only process by which living things are able to synthesize carbohydrates. For instance, many bacteria are able to live by chemosynthetic means. They obtain the energy they require by means of inorganic changes, that is, from reactions of an inorganic oxidation–reduction type, instead of using sunlight as their energy source. But the energy available from inorganic sources is negligible compared with that from sunlight, and photosynthesis is by far the most important process producing the basic foods of the whole biological world.

Position on beech tree	Height from ground	Plant found	Minimum light intensity required to complete the compensation period in 2 h (*lux*)	Optimum intensity required to complete the compensation period (*lux*)	Ecological class of plants
Stump	0.3–0.6 m	Bryophytes, e.g. *Thuidium*, *Hylocomium*, *Thamnium*	400	10000	Shade plants
Trunk	4.8–5.1 m	Bryophytes, e.g. *Anomodon* and some lichens	1200	15000	Intermediate group between shade and sun
Crown	7.9–8.1 m	Mostly lichens, e.g. *Parmelia* and the bryophyte	6000	20000	Sun plants
		Ulota crispula	2000	20000	

Modified *after* Hosokawa and Odani (1957). Compensation period and vertical range of Epiphytes. *Journal of Ecology*, **45**, No. 3, p. 901.

3.2   An outline of the process

Essentially the process involves the combination or fixation of carbon dioxide; hydrogen, released from water, is then used to reduce the combined carbon dioxide so as to form a carbohydrate. Sunlight is necessary to provide the energy for the splitting of water into hydrogen and oxygen. This reaction takes place only in the presence of the green chlorophylls, which act as catalysts, converting light energy into chemical energy. The following equations can be used to summarize these steps:

(i)
$$4H_2O \xrightarrow[\text{green pigments}]{\text{light}} 4(OH) + 4(H)$$

The hydroxyl reforms water, and oxygen is produced:

$$4(OH) \longrightarrow 2H_2O + O_2$$

(ii)
$$CO_2 + 4(H) \longrightarrow (CH_2O) + H_2O$$

In the second equation (CH_2O) is used to represent a simple carbohydrate. From these equations a single general equation is obtained:

(ii)
$$CO_2 + 4H_2O \xrightarrow[\text{green pigments}]{\text{light}} (CH_2O) + 3H_2O + O_2$$

In fact no such simple carbohydrate exists, but hexose sugars ($C_6H_{12}O_6$) and starch are frequently found and so the following general equation is still often used:

(iv)
$$6CO_2 + 6H_2O \xrightarrow[\text{chlorophyll}]{\text{light}} C_6H_{12}O_6 \quad + 6O_2$$

hexose sugar → starch

A series of quite simple tests can be used to verify this equation.

(1) *Necessity for carbon dioxide*

Cover a few leaves of dead-nettle or an ovate-leaved plant for about ten hours so that any starch already in the leaves due to previous photosynthesis may be removed. Fill two large specimen tubes one with $2\,cm^3$ strong potassium hydroxide to remove any carbon dioxide, and the other with $2\,cm^3$ distilled water, or a saturated solution of sodium chloride, as a control. Place one leaf in each tube, so that half the leaf-blade is inside the tube and half outside; carefully cork the tubes (see fig. 3.3). If possible keep the leaves attached to the plant throughout the experiment, otherwise place the tubes in a sealed jar containing a little water; this prevents the leaf from wilting. Illuminate them brightly, preferably with a fluorescent light system, for about two hours, then remove the leaves. Mark with a nick the one that has been in the carbon dioxide-free atmosphere and kill the leaves by immersing them for a few seconds in boiling water. Finally, extract the chlorophyll from the leaf by boiling it for a few minutes in 90 per cent ethanol on a water-bath. Test each leaf for starch with iodine solution; there should be no starch in the part of the leaf that has been in the carbon dioxide-free atmosphere, but both the water control and also the part of the leaf that has been outside the specimen tube should possess starch.

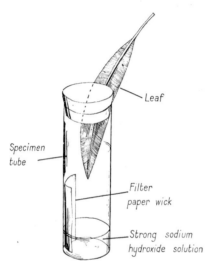

Leaf

Specimen tube

Filter paper wick

Strong sodium hydroxide solution

Fig. 3.3. Device used to show that carbon dioxide is necessary for photosynthesis.

(2) *Necessity for water*

For photosynthesis to take place the plant must be fully hydrated and its cells turgid. There are no simple tests to show how water is utilized.

(3) *Necessity for light*

Keep a *Pelargonium* plant in the dark for about ten hours to remove any starch. Cover part of the leaf with black paper to exclude light and illuminate brightly for about two hours, remove the chlorophyll and test for starch as described in (1) above. Starch should be present only in that part of the leaf that has been exposed to light.

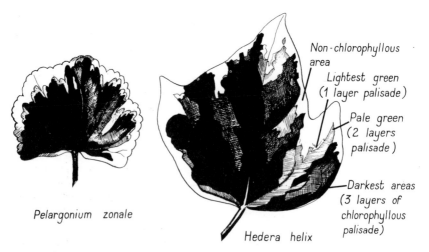

Non-chlorophyllous area

Lightest green (1 layer palisade)

Pale green (2 layers palisade)

Darkest areas (3 layers of chlorophyllous palisade)

Pelargonium zonale

Hedera helix

Fig. 3.4. Variegated leaves of garden geranium (*Pelargonium zonale*) and ivy (*Hedera helix*).

(4) *Necessity for Chlorophyll*

After a period of bright sunlight pick a few young leaves of a variegated plant such as *Acer negundo variegatum*, *Elaeagnus pungens aureovariegata* or a variegated form of *Pelargonium* (see fig. 3.4). Trace the pattern of the variegation, extract the chlorophyll and test for starch in the usual way. Starch should be present only in those parts of the leaf that originally contained chlorophyll.

(5) *The formation of carbohydrates*

The presence of free hexose sugars (e.g. glucose) in leaves that have been photosynthesizing can be shown by Benedict's test (see p. 206). Make a cell extract by grinding in a mortar a few leaves of *Tropaeolum*, *Pelargonium* or onion. Remove most of the chloroplast and cell-wall material by filtering or centrifuging and test the clear liquid for soluble reducing sugar. Quantitative treatment may allow for comparison of the amounts of reducing sugar formed in different plants under a variety of circumstances. The density of the red copper oxide precipitate is a measure of the amount of sugar present.

Although hexose sugars are regarded as the primary product of photosynthesis, they are normally converted to starch rather quickly. As starch is a relatively insoluble carbohydrate, it tends to remain where it is formed, and thus, in many ways it is easier to test for the presence of starch rather than hexose sugars when endeavouring to demonstrate the occurrence of photosynthesis.

(6) *Production of Oxygen*

Keep some fresh sprigs of *Elodea canadensis* under water, with a funnel and test-tube set up to collect any gases evolved. Place in a well-lighted place, and after a reasonable quantity of gas has been collected, test for the presence of oxygen either with a glowing splint or, better, with alkaline benzene-1,2,3-triol (pyrogallol) (see Appendix, pp. 209 and 255).

These experiments are useful in that they illustrate the overall equation, but it has been realized for many years that photosynthesis is a much more complex and many stage process.

3.3 Evidence that photosynthesis is a several stage process

In the 1900s F. F. Blackman investigated the effects of light intensity and temperature on the rate of photosynthesis. He worked with the aquatic willow moss (*Fontinalis antipyretica*), which is particularly useful for such experiments, as the plant possesses no stomata or cuticular thickening, and takes up carbon dioxide readily from its aquatic medium. Similar experiments can be carried out using the aquatic *Elodea canadensis*. Well-illuminated pieces of *Elodea* produce bubbles of a gas, which is mostly oxygen, from their leaves and cut ends of stems (see Appendix, p. 208). The number of bubbles produced from a single stem in a given time gives an indication of the rate of photosynthesis. By varying the distance of the light from the stem it is possible to work out the relationship between light intensity and the rate of the process (see fig. 3.5). The rate of

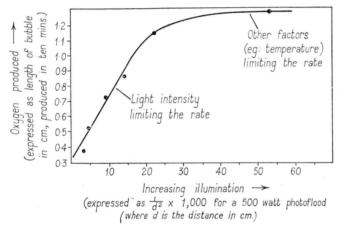

Fig. 3.5. The relationship between light intensity and rate of photosynthesis. (Class result.)

photosynthesis is seen to be directly related to light intensity up to a point; then some other factor, possibly carbon dioxide availability, affects or *limits* the rate of photosynthesis.

A similar experiment can be carried out to determine the effect of temperature. If this is operated at low light intensity, a rise in temperature from 18° to 28°C. makes little difference to the rate of the reaction, but if it is carried out at a high light intensity the rate of photosynthesis is doubled. In the first case light intensity is limiting the rate of the reaction. The second experiment shows that there must be a distinct chemical, non-light requiring stage or stages in photosynthesis as well as a purely photochemical one. Ordinary chemical reactions are strongly influenced by temperature, and for a $10°$ rise in temperature, normally show a doubling in rate. They are then said to have a Q_{10} of two. Light-controlled reactions are temperature insensitive and show no change in rate over the same range in temperature, hence their $Q_{10} = 1$.

Elegant experiments by Warburg in 1919 on the assimilation of carbon dioxide in photosynthesis involved the use of bright flashing lights. Fig. 3.6 illustrates the results of such an experiment. The plant material was illuminated through a rotating disc providing equal durations of light and dark. Slow rotation produced a considerable reduction in assimilation, but flash periods of about seven seconds, although the overall light was still reduced by fifty per cent, caused a reduction of only thirty per cent in the amount of carbon dioxide taken up. Very fast flashes of 3.8 ms caused only a minute reduction in the rate of photosynthesis compared with continuous illumination. This work confirmed that of Blackman in showing that some stages of the photosynthetic process were not light requiring; during the dark period of the flash purely chemical reactions could take place so that reactants were ready to react quickly and effectively as soon as the light was on once again.

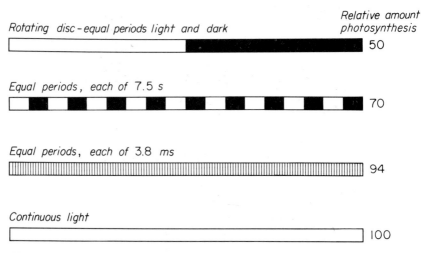

Fig. 3.6. Diagram to illustrate Warburg's Light Flash experiment.

These experiments suggest the existence of three stages in photosynthesis. First, a *diffusion stage*, in which the rate of the process may be governed by the availability of carbon dioxide and the rate with which it can diffuse into the photosynthesizing cell; secondly, a *light-requiring stage*; and thirdly, one or more *chemical stages*. One of these chemical stages, the *combination stage*, in which carbon dioxide is accepted, or fixed, even in the dark, is shown by keeping a plant in the dark in an atmosphere containing carbon dioxide isotopically labelled with [14]C. Autoradiographs produced by placing a photographic plate over the leaves of the plant should show some accumulation of the tracer in the leaves, though there will be less than in those parts exposed to the light (see fig. 3.7 and Appendix, p. 222).

3.4   Detailed examination of the stages of photosynthesis: The diffusion stage

The main factors controlling the diffusion of gases into and out of the leaf are the structure of the leaf itself and also the physical conditions: temperature, humidity, wind velocity and light intensity, which may have direct or indirect effects. A certain amount of carbon dioxide undoubtedly enters the leaf through the upper and lower epidermis, as normal mesophytes have only a fairly thin cuticle overlying the epidermal layers. There is little doubt though that the stomata are normally the chief means by which gaseous exchange takes place. Carbon dioxide uptake is therefore subject to the same controls as water loss, and those factors which cause stomatal closure not only reduce transpiration, but also reduce the rate of photosynthesis. (See discussion of stomata on p. 25.)

Further information about the diffusion stage can be obtained by treating a

Fig. 3.7. Autoradiograph of a shoot of tomato that has been in an atmosphere containing $^{14}CO_2$ for 12 hours. Note that a small amount of tracer is present in areas that have been shielded from the light. Some of this may be due to translocation of substances containing ^{14}C from the illuminated areas. (Marked with an ★.)

leaf so as to prevent gases from entering it. In the following experiment the amount of diffusion through the upper and lower surfaces can be compared.

To investigate the means of entry of carbon dioxide into the leaf

Keep a plant of *Pelargonium* in the dark for twenty-four hours to remove any starch. Then treat as follows four young leaves intact on the plant:

(1) Vaseline both surfaces.
(2) Vaseline the under surface only.
(3) Vaseline the upper surface only.
(4) Leave untreated.

Take care not to damage the leaves when applying the vaseline. Illuminate the leaves with bright fluorescent lamps. After some hours kill the leaves by dipping them in boiling water, extract the chlorophyll in 90 per cent ethanol, boiling on a water-bath, and test for starch with iodine.

As would be expected, most starch accumulates in the untreated leaf and in that which was vaselined only on the upper surface. This experiment emphasizes the importance of the diffusion stage and the structure of the leaf itself in determining the amount of gaseous diffusion, and thus the amount of photosynthesis.

3.5   The combination stage

Studies of the anatomy of leaves, in particular views of the palisade and mesophyll obtained with the scanning electron microscope, have shown that ample air spaces exist throughout the internal tissues of the leaf for carbon dioxide to diffuse to the actual cell where photosynthesis is taking place. At the cell wall the carbon dioxide dissolves in water to form carbonic acid and the carbon dioxide enters the cytoplasm in this form. This is, in effect, a preliminary to the true chemical combination stage as the formation of carbonic acid is easily reversed. The use of radioactive carbon dioxide, $^{14}CO_2$, indicates that even in the dark there is an accumulation of carbon dioxide which does not escape. In other words, the carbon dioxide has somehow become irreversibly fixed into the mesophyll and palisade cells.

A full explanation of the mechanism of fixation has come largely from the use of ^{14}C in conjunction with paper chromatography. Paper chromatography provides us with a method for identifying any intermediates and products of a reaction, while use of tracer carbon can indicate the reaction pathway.

Use of labelled carbon dioxide and paper chromatography in working out the details of the combination stage

The usual materials used in experiments of this kind are unicellular green algae such as *Chlorella* or *Scenedesmus*. These are placed in a large flat-sided flask which can be brightly illuminated and which can be connected to a supply of labelled carbon dioxide. Sampling is carried out by opening the tap on the funnel and allowing some of the algae and their watery medium to pass into a tube containing methanol. This immediately kills them, and an extract of the materials formed in the algae is easily obtained from the alcohol. The apparatus used is illustrated in fig. 3.8. The extract is then examined by the paper chromatographic technique (see Appendix, p. 211).

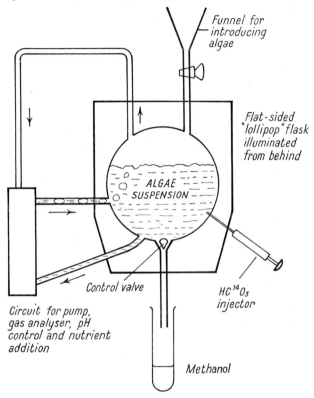

Fig. 3.8. Apparatus used for obtaining extracts of algae that have been photosynthesizing in $^{14}CO_2$. (After Bassham (1962). The Path of Carbon in Photosynthesis. *Scientific American*, **206**, No. 6, p. 88.)

Tracer appearance on the chromatogram is quite easily analysed by placing the tube of a Geiger counter over the various spots on the chromatogram and recording the number of ionizations per second in each case. Alternatively, the chromatogram can be placed on a sheet of photographic negative for a few days. The negative will be exposed where tracer is present in the spot. A version of this technique is described in the Appendix on p. 222.

Results using tracer and chromatography on the dark-fixation or combination stage

If the algae which have been in the dark for some hours and in an atmosphere of $^{14}CO_2$ are analysed as described above, it is found that there is little accumulation of tracer in the algae. If, however, analysis of the algae is done after a preceding light period, applying the $^{14}CO_2$ at the moment the light is switched off, then labelled phosphoglyceric acid (PGA) appears on the chromatogram (other substances such as 2-hydroxybutanedioic acid (malic acid) may be formed by a different pathway, which is discussed on p. 65). This acid is formed as a product of carbon dioxide fixation in photosynthesis. Obviously the carbon dioxide must combine with something to form the three carbon acid, PGA. It is possible that some carbon dioxide acceptor is formed in the light period which is able to combine with the carbon dioxide to form phosphoglyceric acid (PGA) even in the dark, until the acceptor is all used up. The PGA accumulates because some reaction which light usually catalyses can no longer take place in the dark, and this reaction is necessary for the removal of PGA. Figs. 3.9 and 3.10 illustrate these relationships.

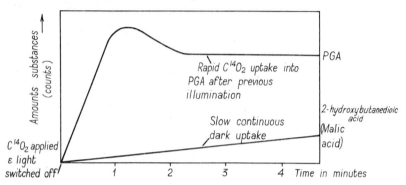

Fig. 3.9. The fixation of $^{14}CO_2$ to form PGA in the dark after a light period. (After Calvin.)

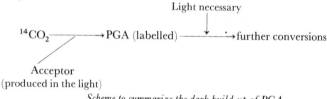

Scheme to summarize the dark build-up of PGA

Identification of the carbon dioxide acceptor

Professor Calvin of the University of California was the first to sort out the sequence of the reactions and recognize the carbon dioxide acceptor. One of the spots formed on his chromatograms, obtained after a long light period, turned out to be a five-carbon sugar, *ribulose-diphosphate* (RDP). Analysis of the levels of RDP along with PGA and other intermediates (see fig. 3.10) showed that the level of RDP fell off sharply in the dark, indicating that it was formed during active photosynthesis and was utilized in a dark reaction resulting in the formation of phosphoglyceric acid. A final and neat check experiment involved the removal of carbon dioxide from actively photosynthesizing *Scenedesmus* (see fig. 3.11). At once the level of PGA fell off while that of RDP rose sharply, as there was no carbon dioxide with which it could combine. When addition of this sugar was found to stimulate the uptake of carbon dioxide and the formation of PGA there was little doubt that this sugar was the carbon dioxide acceptor. Later Calvin was able to identify a specific enzyme, *carboxydismutase*, which catalysed this combination of carbon dioxide with the ribulose-diphosphate. One molecule

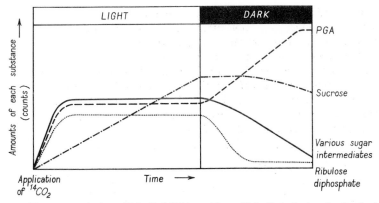

Fig. 3.10. The accumulation of labelled PGA and loss of labelled ribulose in the dark. (After Bassham, *J. Chem. Educ.*, November 1959.)

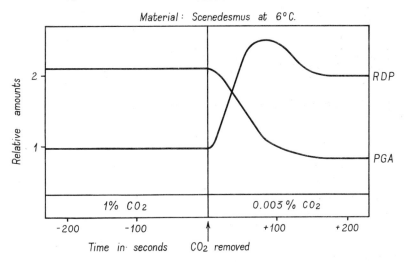

Fig. 3.11. The relative concentrations of PGA and ribulose in the light under different CO^2 concentrations. (After Bassham, *J. Chem. Educ.*, November 1959.)

of this five carbon sugar combines with one molecule of carbon dioxide (in the form of carbonic acid) to form two molecules of PGA. The following scheme shows how this change takes place:

$$
\begin{array}{l}
CH_2O\!-\!P \\
C=O \\
CHOH \\
CHOH \quad +\ H_2CO_3 \longrightarrow \\
CH_2O\!-\!P
\end{array}
\qquad
\begin{array}{l}
CH_2O\!-\!P \\
CHOH \\
COOH \\
CH_2O\!-\!P \\
CHOH \\
COOH
\end{array}
$$

Ribulose-diphosphate + carbonic acid $\longrightarrow$ 2 molecules of phosphoglyceric acid

3.6   The photostage

It was mentioned above that PGA accumulated in the dark, and it was suggested that a light-requiring process took place in normal photosynthesis for the further conversion of PGA. Calvin showed that the conversion of PGA was the only light requiring reaction by sampling the various sugars and possible intermediates after a light followed by a dark period. He found that the concentration levels remained much the same or, more usually, fell off considerably—except the PGA, which accumulated. This was good evidence that there was only one light requiring reaction, otherwise other intermediates would also have accumulated. The graph, fig. 3.10, illustrates this experiment.

The next problem that Calvin dealt with was to find out the substances which were formed by the light reaction. His technique was a subtle one; he allowed the algae to photosynthesize in $^{14}CO_2$ for only a very short length of time and then analysed the results chromatographically. He then lengthened the time very slightly and re-analysed the products. By using a series of different light periods in this way and by comparing the results chromatographically he was able to suggest the possible sequence or pathway for substances produced as a result of the single light-requiring reaction. The initial products of the light reaction involved the three-carbon sugar, triose-phosphate:

$$\text{CH}_2\text{O}\!\!-\!\!\boxed{\text{P}}$$
$$|$$
$$\text{CHOH}$$
$$|$$
$$\text{CHO}$$

(see autoradiographs, fig. 3.12). The further conversions of this three-carbon sugar to six-carbon and other sugars are not light requiring and are described under the final dark stage.

How, then, does light result in the conversion of a three-carbon acid (PGA) to a three-carbon sugar, triose-phosphate? It was stated in the general equation that water combined with carbon dioxide to form hexose, and so far scarcely any

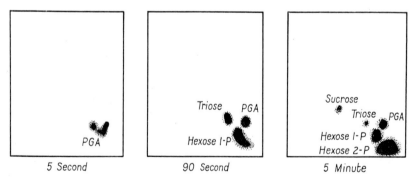

Fig. 3.12. Autoradiographs of chromatograms showing ^{14}C tracer distribution in extracts of algal cells, obtained after various light periods, in $^{14}CO_2$. (After Calvin.)

further mention of water has been made. It has been found that the primary role of light, absorbed by the chloroplasts, is to provide the energy necessary for the splitting of the water molecule into hydrogen and oxygen. The hydrogen is required for the reduction of the carbon dioxide and the oxygen is released.

In 1940 Ruben showed that the oxygen given out in photosynthesis does indeed come from water and not, as was once believed, from carbon dioxide. This was shown by supplying the plant with water enriched with ^{18}O, the heavy but non-radioactive isotope of oxygen. When oxygen released in photosynthesis was examined in the mass spectrometer—an apparatus capable of distinguishing between the isotopic forms—it was found that the gas contained a similar proportion of ^{18}O to the water that was originally supplied. Enrichment of the carbon dioxide produced no corresponding enrichment in the oxygen produced.

Substrate	*Period of* O_2 *collection in min. after start*	*Per cent* ^{18}O *in*		
		H_2O	$HCO_3^- + CO_3 =$	O_2
KHCO$_3$, 0.09 M	0	0.85	0.20	
K$_2$CO$_3$, 0.09 M	45–110	0.85	0.41	0.84
	110–225	0.85	0.55	0.85
	225–350	0.85	0.61	0.86

(After Thomas. *Plant Physiology*. London : J. & A. Churchill.)
Table to show isotopic per cent in oxygen evolved in photosynthesis by *Chlorella*.
(After Ruben et al.)

This was clear evidence that the primary action of light was the splitting of the water molecule into hydrogen and oxygen. Investigation of the photostage must therefore consider both the role of the pigments in absorbing light and also the means by which the energy made available is used to split the water molecule and finally convert the PGA into triose-phosphate.

3.7 The role of the chloroplast pigments

The chloroplasts

We know that the chloroplasts are the site where photosynthesis occurs by examining tissues that have been photosynthesizing for the presence of starch grains; these are found around and inside the chloroplasts. If the saucer-shaped chloroplasts are examined carefully under a high-power microscope the small, round grana of which they are composed may just be visible (figs. 3.13 and 3.15A). Electron-microscope work has revealed that these minute grana are themselves made up of layers rather like a stack of coins (fig. 3.14); these layers are called *lamellae* or *thylakoids* (see fig. 3.15B); there are also lamellae connecting individual grana, the *intergrana lamellae*, the whole forming some kind of loose spiral as is shown in fig. 3.15C. Careful analysis of the individual lamella has indicated that it is composed of layers of chlorophyll and associated pigments separated by lipo-protein membranes (fig. 3.15D and E). Some of the chlorophyll is associated into small groups referred to as *quantasomes*; these are at the outside of the lamella. Each quantasome is thought to be composed of about 230 chlorophyll molecules and it is possible that each is the fundamental light-absorbing unit of the chloroplast. The system in the 4-carbon plants (see p. 65

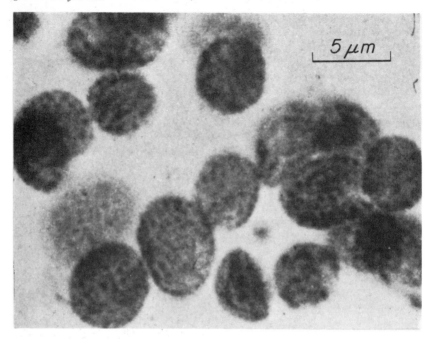

Fig. 3.13. Chloroplasts of broad bean (*Vicia faba*) showing grana. The chloroplast envelope has been stripped off by suspension in hypotonic buffer. (Courtesy of Dr. R. M. Leach, University of York.)

and fig. 3.14) is somewhat different; the grana being much less clear and the chlorophyll arranged in much longer plate-like systems.

Some of the enzymes associated with the process of light energy capture, particularly those concerned with photosynthetic phosphorylation (see p. 62). are found within the grana, while those involved with the chemical stages of the process, the photosynthetic carbon cycle, are situated more in the *stroma* or ground material of the chloroplast. Mitochondria are often found under the saucer-like concavity of the organelle and the formation of starch will usually only take place within the chloroplast if mitochondria are closely adjacent to the membrane surrounding it. This suggests that enzymes or energy in the form of ATP, produced by the mitochondria, are necessary for starch synthesis.

The chloroplast shows a high level of internal specialization and provides a large surface area of pigment capable of absorbing light and converting it to chemical energy.

The pigments

If an extract of the chlorophyll pigments is made in pure propanone and carefully purified it can be analysed by chromatography to identify the pigments found. (For full details of the purification and chromatographic separation of the chlorophyll pigments see Appendix, p. 219.) Figure 3.16 illustrates the resulting chromatogram. The yellow accessory pigment, carotene is least adsorbed and is

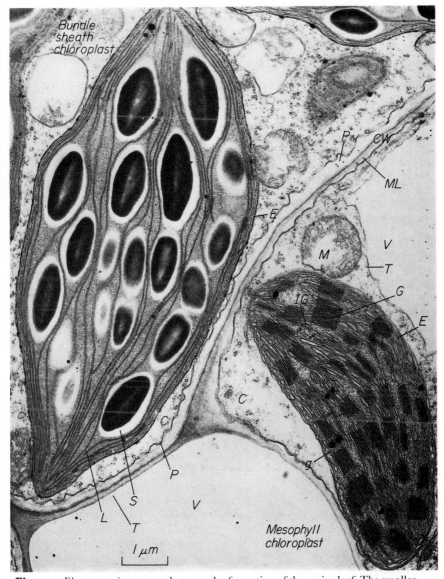

Fig. 3.14. Electron-microscope photograph of a section of the maize leaf. The smaller chloroplast to the right is from the leaf mesophyll, and is typical of the chloroplasts of C-3 plants. That to the left is from the bundle sheath and can carry out the photosynthetic process through the C-4 system. CW=cell wall, with the middle lamella (ML) between the two cells. P=cell membrane or plasmalemma. E=chloroplast envelope. G=granum (a stack-like unit formed where parts of several lamellae lie closely to one another) IG=intergranal lamellae, these are parts of the lamellae lying separately in the stroma. Each lamella consists of a pair of closely set membranes and these are continuous with the granal lamellae. There are no clear grana in the C-4 chloroplasts, the lamellae (L) running as wide plates more or less for the length of the chloroplast.

g=a droplet or globule of lipid in the chloroplast stroma. S=grain of starch.
C=cytoplasm with ribosomes. M=mitochondrion. V=vacuole. (Courtesy of Dr. Jean Whatley and Dr. B. E. Juniper, The Botany School, Oxford.)

A

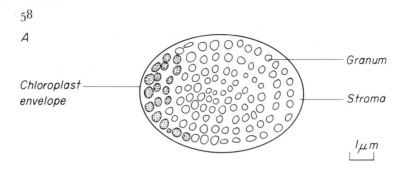

Chloroplast envelope

Granum

Stroma

1 μm

Double membrane or envelope

Lipid droplet

Starch grain

B

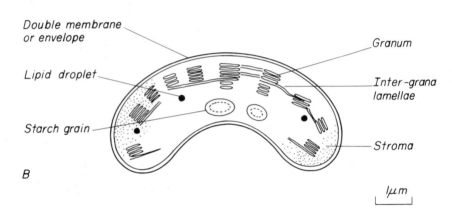

Granum

Inter-grana lamellae

Stroma

1 μm

C

Envelope

Granum

Starch grain

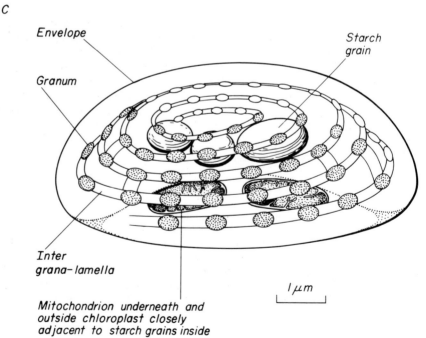

Inter grana-lamella

Mitochondrion underneath and outside chloroplast closely adjacent to starch grains inside

1 μm

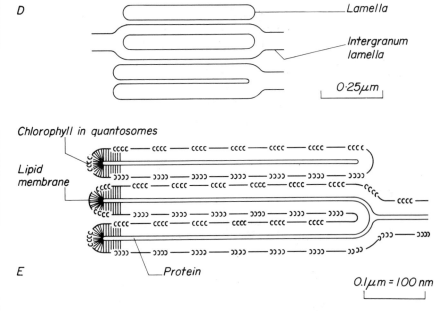

Fig. 3.15. The chloroplast and its grana at different levels of magnification. (A) Surface view of the whole chloroplast. (B) Vertical section through the whole chloroplast. (C) Three-dimensional diagram of the whole chloroplast. (D) Diagram of lamellae within the granum. (E) Interpretation of the lamellae.

found at the top of the paper behind the solvent front. Phaeophytin (grey, a breakdown product of chlorophyll), xanthophyll (yellow-green), chlorophyll *a* (blue-green) and chlorophyll *b* (yellow), as well as other pigments, should also appear. The identity of the pigments can be checked by comparing R_f values with those found in tables. After identification of the pigments by this method it is then possible to separate the pigments in solution so that their light-absorbing properties may be compared. This is best done by using a cellulose column (see fig. 3.17 and Appendix, p. 219). The solutions are rather unstable and should be examined immediately in the spectrometer.

Absorption spectra of the chlorophylls

The spectrometer must first be calibrated; this is most easily done using the mercury-vapour spectrum·(see Appendix, p. 220). The total extracts as well as the separate pigments can then be examined. The results are best recorded graphically, and it is interesting to note how the absorption bands of the individual pigments combine in the absorption spectrum produced from the total extract. Figure 3.18 illustrates the absorption bands with total nettle extract. It should be noted that the absorption bands using petroleum extract are found to be displaced slightly towards the blue end of the spectrum as compared with the spectrum bands obtained by examining the living leaf or a chloroplast suspension.

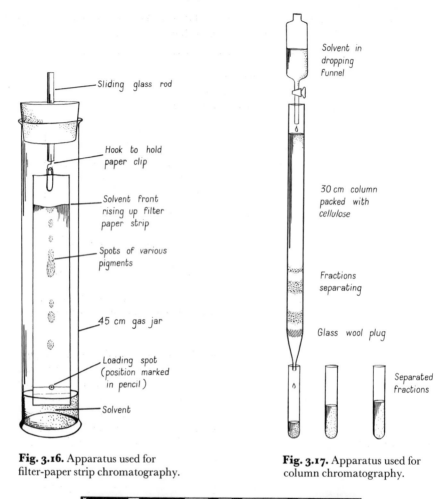

Sliding glass rod

Hook to hold paper clip

Solvent front rising up filter paper strip

Spots of various pigments

45 cm gas jar

Loading spot (position marked in pencil)

Solvent

Fig. 3.16. Apparatus used for filter-paper strip chromatography.

Solvent in dropping funnel

30 cm column packed with cellulose

Fractions separating

Glass wool plug

Separated fractions

Fig. 3.17. Apparatus used for column chromatography.

General extract

RED

750   700   650   600   550   500   450

BLUE

nm

Fig. 3.18. Absorption spectrum of the chlorophyll pigments.

The action spectrum of photosynthesis

Although the general extract of the pigments absorbs light over a wide range of the spectrum, it is not necessarily all utilized in the reactions of the photosynthetic process. Figure 3.19 shows the results when the absorption spectrum and the *action spectrum* of the sea lettuce (*Ulva*) are compared. Near 480 nm absorption is relatively high compared with the amount of photosynthesis occurring; this is due to some absorption by the carotenoids that is not fully utilized in photosynthesis. Techniques of this kind have been of the greatest use in sorting out the roles of various photosynthetic pigments.

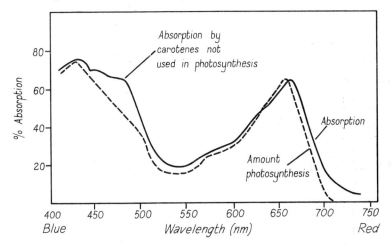

Fig. 3.19. Absorption and action spectra for the sea lettuce (*Ulva lactuca.*) The curves are adjusted so as to correspond at 645 nm. (After Thomas: *Plant Physiology.* London: J. & A. Churchill.)

Important work by Emerson in 1957 showed that if far red light beyond 865 nm was used to illuminate a chloroplast system then relatively little photosynthesis took place, but the yield could be greatly increased by illumination with red of nearer wavelength. The interesting point is that the combined effect of the two qualities of light turned out to be greater than the sum of the two on their own. This *Emerson enhancement effect* therefore gives support to the idea that there are two separate pigment systems operating. These have been called Photo system 1 and 2 and differ as shown in the table below:

	Principle range of absorption (nm)	*Fluorescence*	*Pigments*
Photo System 1	670, 680, <u>700</u> (principally)	Weak	Mostly chloro-phyll *a*
Photo System 2	670, <u>690</u> (principally)	Strong	Mostly chloro-phyll *b* Some chloro-phyll *a*

The Hill reaction

In 1935 Hill, at Cambridge, found that isolated chloroplast suspensions, when illuminated, possessed the power to reduce ferric ions and at the same time produced oxygen. These chloroplast extracts could not utilize carbon dioxide, and in consequence it was assumed that the primary role of the illuminated chloroplasts, on their own, was to split water into oxygen, which was released, and hydrogen. In the *Hill reaction*, as this process has been called, the hydrogen must be accepted by an oxidizing agent; in early work such substances as potassium ferricyanide were used. For instance, the chloroplast extracts of beet and spinach are able to carry out the reaction as follows:

$$4\,Fe^{3+} + 2H_2O \xrightarrow[chloroplasts]{light} 4Fe^{2+} + 4[H]^+ + O_2$$

More recently the redox dye 2:6 dichlorophenolindophenol has been extensively used (see Appendix, p. 221). This is bleached (or reduced) in a few minutes by illuminated chloroplast suspensions, showing that the latter have considerable *reducing power*. In the early 1940's Warburg found that chloride ions had a stimulating effect on the Hill reaction. It has been suggested that they may be important in allowing for the release of oxygen from hydroxyl ions. This may be the reason for the essential requirement of chloride ions by green plants.

Photophosphorylation

The photosynthetic reduction of carbon dioxide is only possible when two conditions are met. First, there must be the kind of system mentioned above to provide reducing power and, second, energy must be made available in the form of ATP (see also p. 85) to drive the whole process.

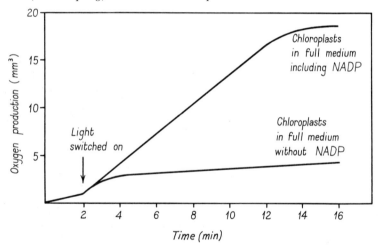

Fig. 3.20. Production of oxygen by chloroplasts in the Hill reaction. Effect of NADP on the rate of oxygen production. (After Arnon (1951). *Nature*, **167**, p. 1008.)

In 1960 D. I. Arnon and his associates at the University of California suggested mechanisms whereby both these requirements could be met simultaneously. They adapted experiments on the Hill reaction to determine whether some of the enzymes and coenzymes that were known to be electron transfer substances in respiration, could also be involved here. It was thought likely that some hydrogen acceptor was able to take up the hydrogen released by the splitting of the water molecule and transfer it to the Calvin cycle (see p. 64). The coenzyme *nicotinamide adenine dinucleotide phosphate* (**NADP**) seemed a likely substance and when this was added to illuminated, isolated chloroplasts a great increase in the rate of oxygen production took place (see fig. 3.20). A number of other redox enzymes and coenzymes have been investigated and two seem to be of particular importance in addition to NADP; these are an iron-containing protein called *ferredoxin* and the cytochromes (see also p. 82). Arnon and others have been able to recognize two different systems both of which, through the mediation of the cytochromes are able to produce ATP from ADP and inorganic phosphate; these are called *cyclic and non-cyclic phosphorylation*. The non-cyclic form will only be described as this also involves the production of reducing power and is the system found in higher plants.

Non-cyclic photophosphorylation

The fact that chlorophyll extracts are highly fluorescent, a feature that can be strikingly demonstrated in ultra-violet light (see appendix p. 221), shows that the molecule is capable of excitation and re-emission of energy. In non-cyclic

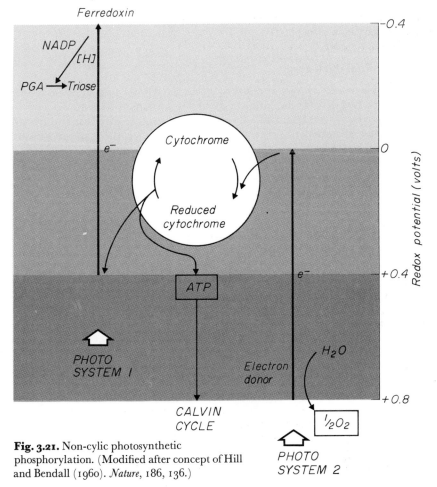

Fig. 3.21. Non-cylic photosynthetic phosphorylation. (Modified after concept of Hill and Bendall (1960). *Nature*, 186, 136.)

photophosphorylation the reasons for the Emerson enhancement effect become apparent. It seems that the Photo System 2 pigments (chlorophyll *b*), absorbing light principally at 690 nm, have their energy status raised so that some of it can be transferred to the second group of pigments, those of Photo System 1 (chlorophyll *a*). These second pigments, in addition to absorbing light at 700 nm can in this way obtain the further boost of energy that is needed for the reduction of ferredoxin. Fluorescence studies with the two pigment systems has shown that chlorophyll *b* fluoresces little in the presence of chlorophyll *a* and this can be taken as evidence that the energy of the chlorophyll *b* is easily transferred to the chlorophyll *a*.

However the energy captured by the Photo System 2 pigments is not wholly transferred to the Photo System 1 pigments as, in the transferance process through the cytochrome system, some energy is made available for the formation of ATP.

The energy changes in the whole system enable water to be split into oxygen, which is released, and the electrons and protons finally enable the reduced ferredoxin to reduce NADP to NADPH. This then completes the sequence by allowing for the PGA–triose conversion. The essentials of these steps are summarized below: (fig. 3.21)

3.8   The final dark stage

There are essentially two sets of conversions involved in the final dark stage. The first is concerned with the formation of the higher carbohydrate products, such as glucose, sucrose and starch, while the second is a series of reactions required to produce the five-carbon sugar, ribulose-diphosphate, the carbon dioxide acceptor which must be regenerated if the photosynthetic carbon cycle is to function. The various sequences of reactions involved here have been worked out by Calvin using the technique described above (see p. 50), in which algae are allowed to photosynthesize for varying light periods in an atmosphere of $^{14}CO_2$.

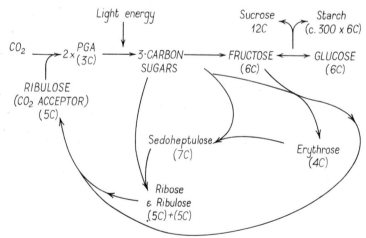

Fig. 3.22. The Calvin cycle. Scheme to show how the various carbohydrate intermediates co-operate in the photosynthetic carbon cycle.

Autoradiographs of the products after periods of five seconds, ninety seconds and five minutes are shown in fig. 3.12, p. 54. These illustrate how, as the PGA (formed in the dark) is used up, first triose-phosphate, then hexose-1-phosphate, and finally sucrose, are formed. These are simplified autoradiographs, but by comparing the accumulation of intermediates in the different chromatograms and also by examining these for their distribution of ^{14}C within the molecule, Calvin was able to show the main steps in both hexose and ribulose-diphosphate formation (see fig. 3.22). It can also be shown that the formation of the higher carbohydrate starch from glucose is possible in the absence of light by a simple experiment (see Appendix, p. 225).

3.9 The C-4 pathway

For several years it was believed that all plants fixed carbon dioxide by means of the Calvin Cycle, but it has recently been found that there are other mechanisms. When $^{14}CO_2$ is fed to the leaves of many tropical grasses, including crop plants such as maize and sugar cane, some species of *Amaranthus* and *Atriplex* as well as a few plants from more temperate regions, for example the cord-grass (*Spartina*), a number of acids are found to incorporate the tracer. The first of these to contain the labelled carbon are the four-carbon acids 2-oxobutanedioic acid (oxaloacetic acid), 2-hydroxybutanedioic acid (malic acid) and aspartic acid; so these species are known as C-4 plants as opposed to the C-3 plants that use the Calvin cycle alone. The primary reaction is between carbon dioxide and phosphoenol pyruvic acid (PEP) to produce 2-oxobutanedioic acid (see fig. 3.23). This acid is

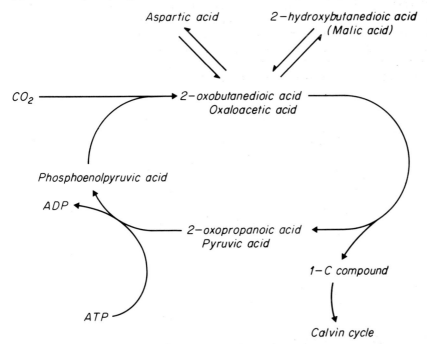

Fig. 3.23. Outline of the C-4 pathway.

then broken down to 2-oxopropanoic acid (pyruvic acid) and to an unknown 1-carbon compound which is then incorporated into the Calvin cycle, by a means yet to be discovered.

C-4 plants not only have a different biochemistry, they also share changes in leaf anatomy, cell structure and physiology. It seems that the C-4 pathway is only found in chloroplasts with a different structure to normal (see fig. 3.14), and that these chloroplasts only occur in specialized mesophyll cells that form sheaths surrounding the vascular bundles. The most important difference between C-3 and C-4 plants is that the former (the majority of plant species) lose somewhere between a quarter and a half of all the CO_2 that is incorporated by photosynthesis by a wasteful process known as photorespiration. Apparently a substantial proportion of the immediate products of photosynthesis is converted to glycollic acid. This is a toxic product which is removed by oxidation, so producing carbon dioxide again. The plant cannot use any of the energy released by this oxidation, so the process is a direct loss to the plant. The reason for the efficiency of the C-4 seems to be at least partly due to a genuine reduction in the rate of glycollic acid formation, but it may also be due to the improved recycling of respiratory CO_2. The enzyme which incorporates CO_2 into the photosynthetic system, phosphoenol pyruvate carboxylase, has a much higher affinity for CO_2 than the corresponding enzyme, carboxydismutase, of the Calvin cycle.

When the overall carbon economy of C-3 and C-4 plants is compared, the latter possess an advantage in that their rates of CO_2 fixation are up to 50 per cent greater under comparable environmental conditions. The C-4 pathway only seems to operate efficiently, however, at fairly high temperatures, so that it is commoner in plants growing in tropical climates. The C-4 pathway occurs in several widely separated groups of species, and yet it is always associated with modified chloroplast structure, situated in specialized bundle sheath cells and with reduced rates of photorespiration. This combination is a remarkable example of parallel evolution in the plant kingdom. On the other hand some quite closely related species may utilize the different systems; for example *Atriplex rosea* is a C-4 plant and *A. patula* a C-3 plant. This suggests that the differences in the metabolic systems may be due to quite small genetic differences which could possibly be bred into species that do not have the more efficient C-4 system at the present time.

3.10   The economics of photosynthesis

In conclusion, it is perhaps worth returning to some of the ecological aspects of the process as photosynthetic capacity is one of the more important factors governing plant distribution. In agriculture and horticulture too, any ways which can be found to increase photosynthetic productivity are bound to be of considerable economic importance.

We know that the biosphere is dependent on photosynthesis to provide the food that living things require for life and it is of course the process which allowed the vast Coal Age forests 300 m years ago to grow and provide us with coal and oil. At the same time photosynthesis maintains the levels of oxygen and carbon dioxide in the atmosphere. Industrial production of carbon dioxide has begun to upset this balance; at the turn of the century the carbon dioxide level in the atmosphere was about 300 p.p.m. (parts per million by volume), at the present time (1977) it is about 320 p.p.m. and it is predicted to rise to over 350 p.p.m. by the end of the century. These changes are bound to influence the biological productivity of the plant kingdom as carbon dioxide concentration is usually the most serious limiting factor to the process of photosynthesis. On a restricted scale it is now common commercial practice to raise the level of carbon dioxide in the atmosphere of greenhouses artifically. This is usually done in the winter months by burning propane or relatively pure hydrocarbon fuel so that the air is both warmed and enriched with carbon dioxide.

Over the years farmers and horticulturalists have endeavoured to select strains of crops which are particularly effective in both light capture and conversion to stored product. Even so the efficiency of light capture is not high; a plant in full summer sunlight receives between five and ten times its saturating light intensity, and is therefore inefficient in light use. More precise information is available on the synthesis of organic matter by crop plants. High yielding varieties of cereals in Britain have recently produced as much as 7 tonnes of harvested organic matter per hectare. The harvested crop, however, is only a part of the total, and the biological productivity would be around 18 tonnes per hectare. This would have an energy content of c. 3×10^5 MJ. The annual incident visible radiation in Sourthern England is c. 1.4×10^7 MJ ha^{-1} so that the crop growth represents a light conversion of around 2 per cent. It must be remembered, however, that this calculation was performed on an exceptionally high yielding crop, and that the average crop yields, on both a national and world basis are substantially less than this. Thus the efficiency of photosynthesis as a means of capturing and utilizing sunlight is very low.

Up to the present time most work on selection of high-yielding crops has been done with regard to such features as yield, rate of growth, soil requirements, hardiness and disease resistance in mind. Recent work on C-4 plants suggests that the advances of the next decade may involve selection for more physiological adaptations.

Further reading on Photosynthesis

BASSHAM, J. A. and CALVIN, M. (1957). *The Path of Carbon in Photosynthesis*. Englewood Cliffs, Prentice-Hall.

FOGG, G. E. (1968). *Photosynthesis*. London, English Universities Press.

HALL, D. O. and RAO, K. K. (1977). *Photosynthesis*, 2nd edn. Institute of Biology's Studies in Biology no. 37. London, Edward Arnold.

TRIBE, M. A. and WHITTAKER, P. A. (1972). *Chloroplasts and Mitochondria*. Institute of Biology's Studies in Biology no. 31. London, Edward Arnold.

WHITTINGHAM, C. P. (1971). *Photosynthesis*. Oxford Biology Reader No. 9. Oxford University Press.

4

Respiration

4.1 The process

In the last chapter we dealt with the process by which green plants manufacture various carbohydrate food substances, incorporating the sun's energy into the plant. Respiration is a reversal of photosynthesis in that it is the means by which the food materials, such as glucose and fructose, are broken down into simpler substances, releasing energy in the process. The energy released is hardly evident externally (except in the form of some heat), as it is channelled into other systems and finally used for the various energy-requiring reactions going on within the plant. As no organism can do without energy, which it needs for these general processes of metabolism, respiration, like photosynthesis, is a vital process in living organisms.

By measurement of the volumes of the gases required and evolved and by the study of the conditions and requirements of the process the following general equation for the breakdown of hexose sugar is often written:

$$6O_2 + C_6H_{12}O_6 = 6CO_2 + 6H_2O$$

This kind of respiration, utilizing oxygen, is the means by which most plants and animals obtain a great part of their energy, and is called *aerobic respiration*. On the other hand, many plant tissues under conditions of low oxygen availability obtain their energy *anaerobically*. The following equation summarizes this process:

$$C_6H_{12}O_6 = 2CO_2 + 2C_2H_5OH$$

Instead of complete breakdown of the hexose to carbon dioxide and water taking place, ethanol and carbon dioxide are formed. Plants are unable to utilize this ethanol, which therefore represents a considerable wastage of energy, aerobic respiration being more than twelve times more efficient a process. In addition, small concentrations of alcohol are toxic and so most plants are unable to tolerate anaerobic conditions for long periods.

Extensive research using the techniques of chromatography, autoradiography and inhibitor treatments have revealed that respiration, like photosynthesis, is a many-stage process, and the general equations given above are simply extreme summaries of what goes on. Work has shown that aerobic and anaerobic respiration share a common, non-oxygen requiring pathway called *glycolysis* in their first stages, that is, in the breakdown of hexose into the simple three-carbon acid, 2-oxopropanoic acid (pyruvic acid, $CH_3.CO.COOH$). If oxygen is present, then the 2-oxopropanoic acid is utilized through a series of conversions involving several tricarboxylic acids; this sequence is known as the *Krebs' Cycle*.

At various stages during the cycle oxygen is required and water and carbon dioxide released, energy being made available in the process. On the other hand, if there is no oxygen a 'switch' occurs and the anaerobic sequence takes place, resulting in the production of alcohol. These stages are summarized in fig. 4.1.

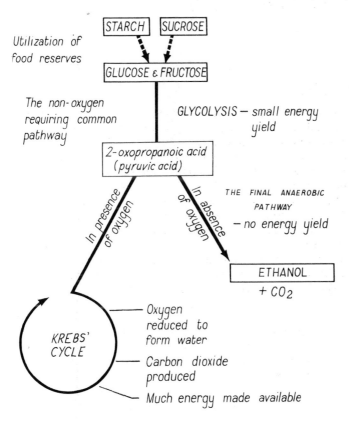

Fig. 4.1. The main stages of respiration.

In considering respiration there are two main lines of approach; first, that involving the effect of various factors on the rate of respiration of whole tissues or intact organisms, and secondly, that involving extraction of the individual substances concerned in the process and investigation of their properties; these are the biochemical aspects.

4.2 Measurement of the rate of respiration

A considerable amount can be learned about both the biology of the intact plant and the nature of the respiratory process itself through studies of the rate of respiration. The rate can be measured either in terms of oxygen uptake or of carbon dioxide production.

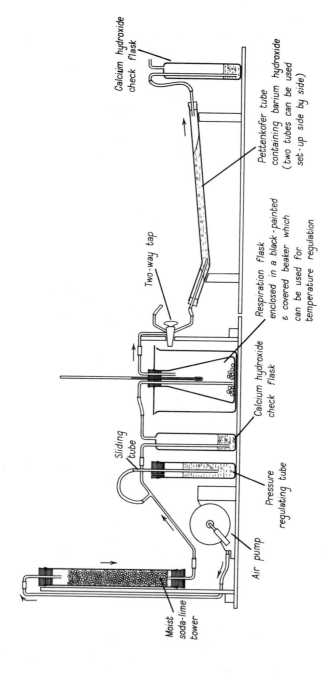

Fig. 4.2. The Pettenkofer apparatus. Used for the quantitative estimation of the carbon dioxide produced in respiration.

Measurement of carbon dioxide production

The Pettenkofer technique (see fig. 4.2) is used to measure the quantity of carbon dioxide produced by a given weight or volume of material in a given time. Essentially, carbon-dioxide-free air is supplied to the respiring material and the respiratory gases passed into baryta (barium hydroxide). The carbon dioxide reacts with the baryta to form insoluble barium carbonate. After a given period the unchanged baryta is titrated against standard hydrochloric acid using phenolphthalein as indicator to find how much baryta is left, and hence how

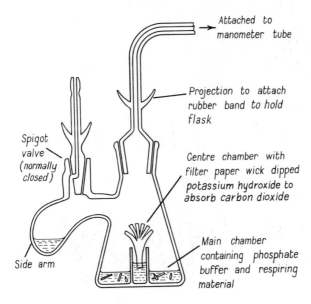

Attached to manometer tube

Projection to attach rubber band to hold flask

Spigot valve (normally closed)

Centre chamber with filter paper wick dipped potassium hydroxide to absorb carbon dioxide

Main chamber containing phosphate buffer and respiring material

Side arm

Fig. 4.3. The flask of the Warburg manometer.

much of the baryta has been used up by reaction with the carbon dioxide. From this value the amount of carbon dioxide produced per unit of respiring material per hour can be calculated (see Appendix, p. 226).

The Pettenkofer apparatus can be used to compare the rate of respiration of various whole plants, or to investigate changes in respiratory rate during plant growth or senescence, for instance, during germination. The apparatus can also be used to measure the effect of various factors, such as temperature or oxygen availability, on the rate of the process.

Measurement of oxygen uptake

The Warburg manometric method is the most useful for determining the rate of oxygen uptake. In this device (see figs. 4.3, and 4.4) the respiring material is enclosed in a small flask and kept at a constant temperature in a thermostatically controlled water-bath. A small manometer is attached to the flask, and this registers any gases produced or taken up. Potassium hydroxide can be included in the centre well to absorb carbon dioxide as it is produced so that the apparatus is able to record the amount of oxygen utilized. The flask is also provided with a side arm from which inhibitors or special substrates can be added once the normal rate of respiration of the material has been determined. A blank, control manometer, the thermobarometer, is also run alongside to make correction for any temperature or barometric change. For full details of the technique see Appendix on p. 228.

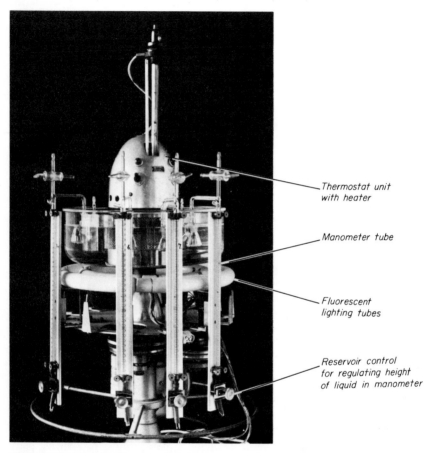

Thermostat unit
with heater

Manometer tube

Fluorescent
lighting tubes

Reservoir control
for regulating height
of liquid in manometer

Fig. 4.4. The Warburg manometer. This device is used for the estimation of gas exchange by respiring or photosynthesizing tissues. The apparatus consists of a temperature-controlled water-bath together with a set of three or more flask and manometer units. The flasks are submerged in the tank and kept shaken by a motor. Light can be provided by a fluorescent tube below the water bath. (Photograph courtesy of John Marshall, Wye College.)

Respiratory quotient

Some information about the substrates being used in respiration and also the type of respiration in progress may be obtained by comparing the amount of carbon dioxide produced with oxygen utilized. The respiratory quotient is the amount of carbon dioxide produced divided by the amount of oxygen utilized over a given period of time.

$$RQ = \frac{\text{Volume } CO_2 \text{ produced}}{\text{Volume } O_2 \text{ absorbed}}$$

RQs for the complete oxidation of various respirable substrates, which can be

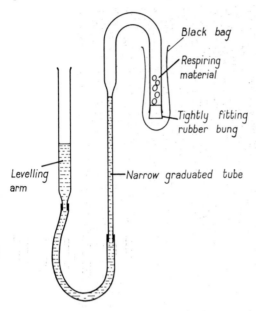

Fig. 4.5. A simple Ganong respirometer. Three sets of apparatus are needed:
(1) Peas+potassium hydroxide in the tube, to absorb carbon dioxide
(2) Peas+water in the tube, to estimate the difference in volume between the gases absorbed and released
(3) Glass beads+water in the tube, to allow for temperature and barometric changes

The calculation of the respiratory quotient (RQ) is as follows. The tubes were set up, levelled and left for 48 hours before re-levelling and reading off the new volumes:

Tube	Contents	Change in volume (cm³)	Corrected for thermobarometer (cm³)	Deduction
1	Peas+KOH	−10.0	−8.0	8 cm³ oxygen consumed
2	Peas+water	+ 0.5	+2.5	2.5 cm³ carbon dioxide produced in excess of oxygen consumed
3	Beads+water (thermobarometer)	− 2.0	—	

Carbon dioxide produced $= 8.0 + 2.5 = 10.5 \text{ cm}^3$

$$RQ = \frac{CO_2}{O_2} = \frac{10.5}{8.0} = 1.33$$

calculated from the equation for the complete oxidation of the substance concerned to carbon dioxide, are given in the table below:

	RQ	Material
	0.7	Fats
	1.0	Carbohydrates
About	0.9	Proteins and amino-acids

In practice, these values are seldom obtained exactly, as complete oxidation rarely occurs and there may be a mixture of substrates being utilized. It can be demonstrated very simply that the RQ of normal germinating peas (which have a food reserve composed largely of carbohydrates) is near unity by placing the peas in the closed end of an inverted U-tube, the other, open, end of which is in

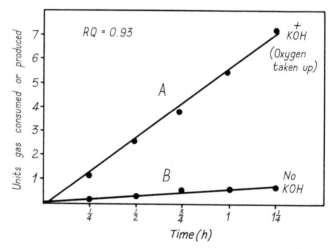

Fig. 4.6. Manometric estimation of respiratory quotient of barley root tips.

A = oxygen taken up
B = excess oxygen taken up over carbon dioxide produced

$$RQ = \frac{vol.\ CO_2}{vol.\ O_2} = \frac{7.0-0.5}{7.0} = \frac{6.5}{7.0} = 0.93 \quad \text{(Class result.)}$$

water. If the peas are left for some time there should be no change in the level of the water in the tube. But if the open end of the tube is placed in potassium hydroxide the liquid level will rise inside the tube, indicating absorption of carbon dioxide. Thus, as there is no change in level in the first case, in spite of carbon dioxide being produced, then the amount of carbon dioxide given out must be the same as the amount of oxygen utilized and the RQ = 1.

If the respiration chamber is connected to a manometer or volume-measuring device, then the RQ can be measured accurately, and this is the principle used in manometric experiments (figs. 4.4 and 4.6) and also in the Ganong respirometer (see fig. 4.5 and Appendix, p. 227).

Respiratory quotients may also give some indication of the type of respiration going on. Clearly an RQ of 1 suggests the fully aerobic respiration of carbohydrates. High RQ values between 2 and 7 are found when the plant is respiring under water-logged, oxygen-deficient conditions. These suggest the occurrence of anaerobic respiration. Low RQs below 0.5 may indicate that some of the carbon dioxide is being fixed, perhaps to form plant acids (e.g. 2-hydroxybutanedioic (malic acid), see p. 65).

Production of heat and energy

The main production of energy in the respiratory process is utilized for the formation of high-energy phosphate, ATP, from ADP (see p. 85). In addition to this, some heat production, which is necessary to maintain a thermodynamic equilibrium, normally occurs, and this may be used to give some indication of the rate of respiration.

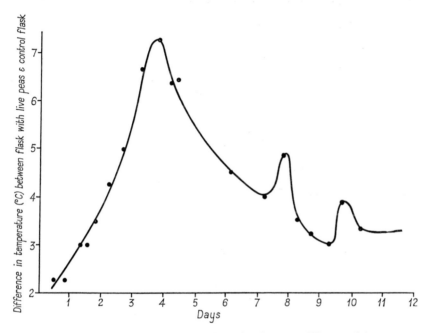

Fig. 4.7. Fluctuations in heat production of germinating peas. (Class result.)

Temperature changes can be detected by placing the respiring material, such as germinating peas, in a thermos flask stoppered with cotton-wool and including a thermometer. A similar control can be set up using germinating peas that have been killed with boiling water and kept sterile by the addition of mercuric chloride to prevent bacterial infection. It is also best to surface sterilize the living peas to kill any bacteria and fungi that may be present. This is best carried out by soaking the peas in a 1 per cent solution of sodium hypochlorite for three minutes and then washing them in sterile distilled water. If the daily measurements of temperature are plotted on a graph the results illustrated in fig. 4.7 may be obtained. This illustrates that the amount of heat evolved rises as germination begins and does not fall off until most of the available substrates have been utilized.

4.3   The effect of various factors on the rate of respiration of the intact organism

Effect of oxygen concentration

The ability of different organisms to survive in different oxygen concentrations can be examined using the Pettenkofer technique. This is most simply carried out by providing the plant with nitrogen and oxygen from cylinders. Different oxygen concentrations can be provided by regulating the oxygen flow compared

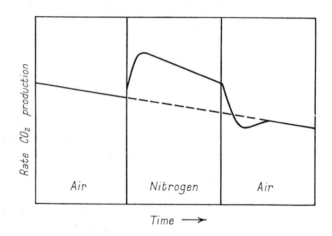

Fig. 4.8. The effect of nitrogen on the respiratory rate of apples. (After Blackman *Analytic Studies on Plant Respiration*. Cambridge, University Press.)

with the nitrogen flow. A pressure-regulating device, as used in the Pettenkofer set-up, should be fitted to both oxygen and nitrogen lines. For short periods the oxygen may be removed by passing the gas through two or three bottles containing alkaline benzene-1,2,3-triol (pyrogallol).

On removal of oxygen the rate of respiration of the organism might be expected to fall off, but in most cases the rate, in terms of carbon dioxide production, actually increases. This indicates that anaerobic respiration comes into action when oxygen is no longer available and that the plant, if it is to make up for the relative inefficiency of this system, has to respire faster. The graph (fig. 4.8) illustrates this effect when apples are in air and nitrogen.

Unless the organism is particularly well adapted to life under low-oxygen concentrations, such treatment will in most cases probably result in the eventual death of the organism. Under natural conditions most plants are aerobic, but even in these there may be times, for instance in the root, when the soil is temporarily waterlogged, and some tissues respire anaerobically. Such organisms are referred to as facultative anaerobes, at least for the tissue concerned. Few plants are able to respire anaerobically for any length of time, a notable exception being the seedling of a rice plant, which is able to survive under completely anaerobic conditions for a considerable period. How far the roots of many marsh and semi-aquatic plants, although they often possess porous, *aerenchymatous* stems and roots, are able to stand up to anaerobic conditions is still largely unknown, and it is interesting to speculate whether such tolerances may be important in determining the competition and subsequent survival of plants in such communities. Better-known facultative anaerobes, such as yeast (*Saccharomyces cerevisiae*), can usually tolerate anaerobic conditions until the accumulation of the poisonous waste product ethanol, which is toxic at 12 per cent concentration, renders a return to aerobic conditions a necessity if the organism is to continue to live. Selection through the ages of good fermenting strains of yeast have enhanced its tolerance for alcohol, but it must be emphasized that in spite of this, it grows much better under aerobic conditions. A similar situation exists in the potato, but lactic (2-hydroxypropanoic) acid is the waste product that accumulates. A few organisms, principally soil bacteria such as the *Clostridia*, are obligate anaerobes and can live in wet, clay soils.

Effect of substrate availability

If apples are placed in the respiratory chamber of the Pettenkofer apparatus and the amount of carbon dioxide produced per day measured over a period of months, then as the apples ripen and different substrates for respiration are made available, the respiratory rate changes.

Such an experiment is clearly rather laborious, but it does give general information on the substrates utilized in apple respiration. Simpler experiments can be done using fungi which grow quickly and can easily be cultured on a variety of substrates. The respiratory rate of colonies of similar size of the same fungus growing on different media, such as lignin, cellulose, starch and glucose, can be found. Investigations on these lines may be of considerable value in ecological studies.

Effect of temperature

Like most chemical reactions, the rate of respiration is influenced by temperature; the higher the temperature, the faster the rate. For instance, estimation of Q_{10} of the process (see above p. 48) for a rise in temperature from 8° to 18°C. gives a Q_{10} of 2, indicating a chemical reaction. But if the rise is at a much higher starting temperature, say between 20° and 30°C., then the Q_{10} may fall below 2. This is partly accounted for by the steps of respiration taking place so fast that the rate falls off, as the oxygen cannot diffuse into the tissue fast enough. In addition, substrate interconversions may take place at higher temperatures. For instance, fats may be formed from carbohydrates by a reaction in which carbon dioxide is utilized and oxygen produced. This will clearly upset the readings for the amount of respiratory carbon dioxide actually produced.

The effect of temperature on different plants is an obvious factor in determining plant distribution. Some plants are better adapted to life at higher temperature than others, and when growing in their most suitable temperature range are best able to compete with their neighbours. This is well illustrated by reference to the temperature required for germination of various seeds. A temperate mesophyte, hairy willow herb (*Epilobium hirsutum*), germinates best at about 21°C., wheat will germinate between 5° and about 34°C., while subtropical maize has a still higher temperature range.

4.4   Biochemical aspects: the anaerobic pathway

(i) *Glycolysis*

The ease with which most plants are able to switch from aerobic to anaerobic respiration and back is perhaps an indication that the processes are very closely related to one another in some of their stages. That they do indeed share a common pathway in part is shown by two main pieces of evidence. First, that the temperature coefficients or Q_{10}s for aerobic and anaerobic respiration are the same, indicating participation by some of the same enzymes at an early stage in both processes, and secondly, the use of the inhibitor sodium iodoacetate has shown that by blocking one reaction it is possible to inhibit both aerobic and anaerobic respiration. This substance inhibits the conversion of triose-phosphate into phosphoglyceric acid by the enzyme triose-phosphate dehydrogenase, which is an early stage in glycolysis. Other substances can be used to inhibit either aerobic or anaerobic respiration separately, which shows that in some stages at least the two processes must be quite different. This primary, anaerobic phase involves the breakdown of hexose sugar, in direct reversal of the photosynthetic sequence, through three-carbon sugars, including phosphogly-ceraldehyde, to phosphoglyceric acid and finally 2-oxopropanoic acid (see fig. 4.9). The individual steps of this process have been investigated by chromatographic and general analysis methods.

An important feature of glycolysis is that two hydrogen atoms must be removed. These are passed to a hydrogen acceptor, NAD (nicotinamide adenine dinucleotide phosphate), which becomes reduced, $NADPH_2$. It is probable that this reaction is coupled with the synthesis of ATP from ADP. Two molecules of ATP are synthesized from ADP for every molecule of hexose sugar broken down; this represents a small release of energy which can be utilized by the plant.

(ii) *The final anaerobic pathway*

In the absence of oxygen the 2-oxopropanoic acid, formed in glycolysis, is broken down to form ethanal (acetaldehyde) and finally ethanol. Carbon dioxide is produced in the first step, and is that which appears during fermentation process:

$$
\begin{array}{ccccc}
CH_3 & & CH_3 & & CH_3 \\
| & & | & +2[H] & | \\
CO & \longrightarrow & CHO & \longrightarrow & CH_2 \\
| & \searrow & & & | \\
COOH & CO_2 & & & OH \\
\text{2-oxopropanoic acid} & & \text{ethanal} & & \text{ethanol} \\
\text{(pyruvic acid)} & & \text{(acetaldehyde)} & &
\end{array}
$$

The Conway method (see Appendix, p. 231) can be used to demonstrate part of this pathway. If an active strain of yeast that has been living under aerobic conditions is fed with glucose in a closed Conway unit, ethanal is formed which diffuses into the centre well of the flask, where its presence is detected by a dilute solution of 2,4-dinitrophenylhydrazine. The formation of ethanol can, of course, be demonstrated by distilling the product of anaerobic respiration.

Although no ATP is formed during this final anaerobic pathway, the importance of the reactions is that they allow for the acceptance of hydrogen. The reduced $NADPH_2$, formed during glycolysis, can be reoxidized by passing the hydrogen to the ethanal which becomes reduced to ethanol. In this way the $NADPH_2$ is made re-available for more glycolysis.

The net gain in ATP per molecule of hexose respired during the whole of anaerobic respiration is thus the two molecules of ATP formed during glycolysis.

4.5 Aerobic respiration: the Krebs' Cycle

Analysis of acids

If some leaves that have been respiring in air are ground up they are usually found to produce an acid reaction when tested with indicators. Some of these acids can be found by means of chemical tests (see Appendix, p. 231). Careful chromatographic analysis will reveal the presence of a wide range of acids. These are mostly constituents of the *Krebs' Cycle* (see fig. 4.9), in which 2-oxopropanoic acid is broken down in the presence of oxygen to yield carbon dioxide and water and bring about the formation of the main part of the plant's high-energy phosphate, ATP.

The evidence that has been used to show how a few steps of the cycle occur is discussed in the following sections.

Use of inhibitors

Inhibitor treatments have proved most useful in sorting out the individual steps in reaction pathways. In general, the principle behind the use of these substances is as follows. Suppose we have a reaction sequence which we believe, from other evidence (e.g. analysis of intermediates), to go $A \rightarrow B \rightarrow C \rightarrow D$ and we think that the reaction $B \rightarrow C$ is controlled by an enzyme for which there is a known inhibitor. Then if we examine and compare by analysis, the occurrence of these substances A, B, C and D before and after inhibition, the inhibited reaction should show much the same concentration of A, more than normal of B and the near absence of C and D. We would obtain confirmation that the reaction $B \rightarrow C$ has been inhibited by adding C and then detecting the additional presence of D.

One of the best examples of an inhibitor that has been useful in the study of the Krebs' Cycle is propanedioic acid (malonic acid). This blocks the conversion of butanedioic (succinic acid) into *trans*-butanedioic acid (fumaric acid). This occurs because propanedioic acid has a structure which is superficially similar to that of butanedioic acid, so that it competes for reaction with the enzyme, butanedioic dehydrogenase, and the normal conversion of butanedioic acid into *trans*-butanedioic acid is slowed up. It is likely that the enzyme forms a complex with the propanedioic acid substrate; but due to the difficulty of removing hydrogen from this substance, it is unable to act and remains as a propanedioic acid–enzyme complex that effectively prevents the enzyme acting in the normal butanedioic–*trans*-butanedioic acid conversion.

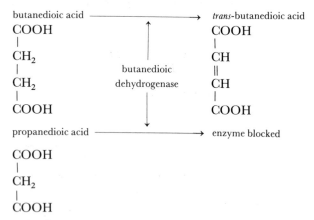

In the presence of propanedioic acid, butanedioic acid accumulates and the concentration of *trans*-butanedioic acid falls off—thus indicating one step in the cycle of acids.

Another inhibitor that has proved useful in fluoroethanoic acid (fluoroacetic acid), for this blocks the enzyme aconitase which is required in the conversion of 2-hydroxypropane-1,2,3-tricarboxylic acid (citric acid) into 1-hydroxypropane-1,2,3-tricarboxylic acid (*iso*citric acid). Fluoroethanoic acid has a structure that is similar to ethanoic acid (acetic acid) except that it has one hydrogen atom of the methyl group replaced by fluorine. It is converted into 1-fluoro-2-hydroxypropane-1,2,3-tricarboxylic acid (fluorocitric acid) by the condensing enzyme which would normally function in the conversion of 2-oxobutanedioic acid (oxaloacetic acid) and acetyl coenzyme A (an 'acetic acid compound') into 2-hydroxypropane-1,2,3-tricarboxylic acid. The enzyme aconitase, which would, in the normal cell, convert the 2-hydroxypropane-1,2,3-tricarboxylic acid into 1-hydroxypropane-1,2,3-tricarboxylic acid, is now unable to utilize 1-fluoro-2-hydroxypropane-1,2,3-tricarboxylic acid, and thus an inhibition of respiration follows. Through the accumulation of 1-fluoro-2-hydroxypropane-1,2,3-tricarboxylic acid by inhibition another step in the Krebs' Cycle has been indicated. It is also interesting, as it illustrates the specificity of an enzyme; in the first reaction the enzyme has a wide specificity and is unable to distinguish between ethanoic and fluoroethanoic acid, but the second enzyme (aconitase) is more specific, and no conversion of alien substance follows. These relationships are summarized in fig. 4.9, also at top of page 82.

Use of radioactive tracers

A good example of the use of radioactive tracers in the study of aerobic respiration comes from feeding tissues with labelled 2-oxopropanoic acid. If the carbon atom of the carboxyl group is labelled with ^{14}C, then no tracer appears in the acids of the Krebs' Cycle when they are examined by chromatograms and autoradiographs. On the other hand, the carbon dioxide produced is highly radioactive. Labelling in the methyl or carbonyl group results in a considerable accumulation of tracer in the acids of the cycle and relatively little in the carbon dioxide. This is direct evidence that the first step in the cycle is one which involves carbon dioxide production as well as indicating that 2-oxopropanoic acid is the starting-point of the cycle of acids.

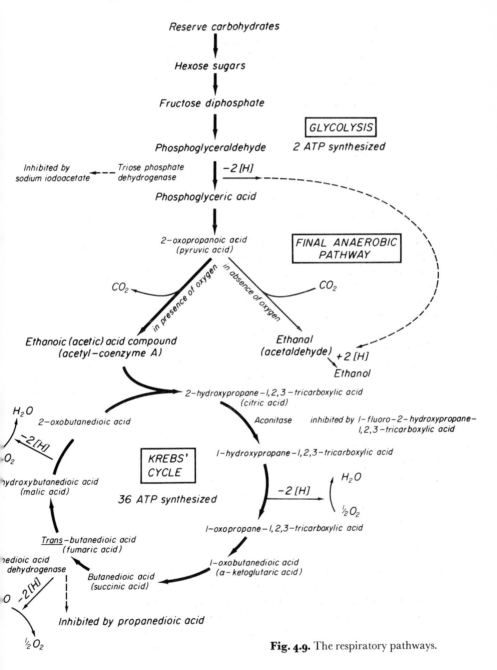

Fig. 4.9. The respiratory pathways.

82

(*a*) *In the normal cell*

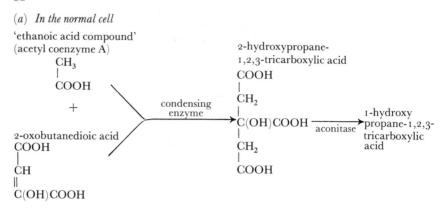

(*b*) *With the inhibitor*

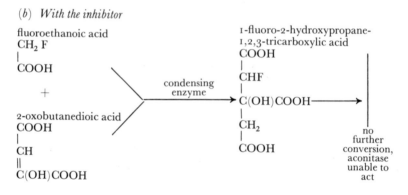

Addition of labelled acids of the cycle and detection of their conversion into other acids has also proved a fruitful line of research, and by the use of tracers and inhibitors, chromatographic and autoradiographic means, the main steps of the cycle have been worked out; 2-oxopropanoic acid is seen to be completely oxidized—that is, all its hydrogen atoms are removed, 2-oxobutanedioic acid being successively used and regenerated.

4.6 The oxidase systems

The next problem is how are these hydrogen atoms removed and the oxygen utilized? Another inhibitor, perhaps the best known of all, potassium cyanide, has been found useful here; it considerably reduces aerobic respiration (see fig. 4.10) by blocking the oxidase enzymes that are responsible for transferring hydrogen to atmospheric oxygen. These oxidase enzymes are part of *redox-chains*, oxidation–reduction systems, in which not only the hydrogen but also electrons are transferred from a substrate so that the reduction of oxygen to form water may ultimately take place. These chains have a number of stages. First, an enzyme (a dehydrogenase) removes hydrogen and an electron from the substrate, secondly, a non-protein *coenzyme* accepts these so that the enzyme is oxidized back to its original state, ready to react with more substrate. At the next stage of the redox-chain the electron causes the reduction of a *carrier enzyme* and the hydrogen ion may be released to the surroundings. This carrier enzyme then

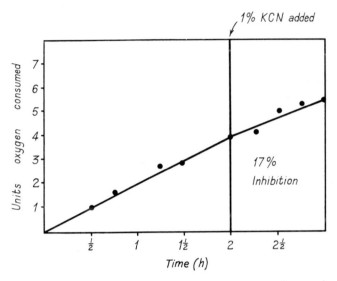

Fig. 4.10. The effect of KCN on the respiration of barley root tips. (Class result.)

transfers its electron to a *terminal oxidase* which then reduces oxygen, so that, with the released hydrogen ion, water is formed. These steps can be summarized as follows:

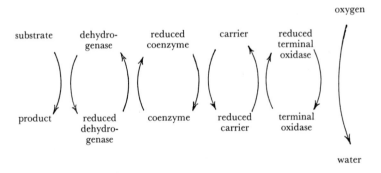

Common coenzymes include NAD (nicotinamide adenine dinucleotide) and NADP (nicotinamide adenine dinucleotide phosphate); cytochrome *c* and ascorbic acid (vitamin C) are common carriers. Some terminal oxidases are cytochrome oxidase, ascorbic oxidase and polyphenol oxidase. A good example of a redox-chain occurs in conjunction with the butanedioic acid to *trans*-butanedioic acid conversion, a reaction which requires the removal of two hydrogen ions and two electrons.* In this case there is no coenzyme and the

* See Appendix, p. 233 for experiments.

dehydrogenase is able to pass its electrons directly to cytochrome *c*.

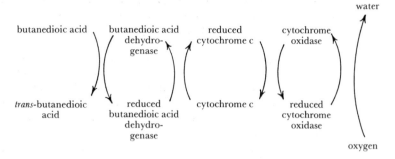

Inhibition by cyanide is usually indicative of the presence of a metallic group in the enzyme concerned, and in the terminal oxidases it is either copper or iron. Cyanide forms stable complexes with thse metals, thus blocking the enzyme. In the cytochromes iron is the metal concerned in the electron transfer system; this can be shown by the use of carbon monoxide, which inhibits respiration in the dark but not in the light. This is because only iron–carbon monoxide complexes are light sensitive. The ascorbic acid and polyphenol oxidase systems contain copper as their metallic or activator group. It is interesting that these transition elements copper and iron play such an important part in respiration, the enzymes concerned depending on the ability of these metals to become successively oxidized and reduced.

The full equation for the conversion of butanedioic acid into *trans*-butanedioic acid would therefore take place as follows:

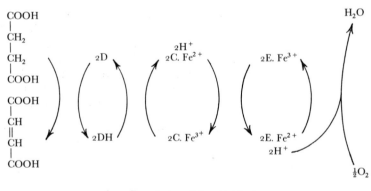

where D = butanedioic acid dehydrogenase
C = cytochrome c
E = cytochrome oxidase

Investigation of the cytochromes

We have seen above that the cytochromes and other oxidase systems have a vital position in respiration, and an understanding of how they occur and function is an important aspect of respiratory studies. If the extracts prepared from young and actively growing plants are examined spectroscopically (yeasts, onions and shallots make particularly good sources, see Appendix, p. 235) it is usual to find a series of weak absorption bands. Under reducing conditions these may be intensified, and under oxidizing conditions they nearly disappear. These absorption bands (see fig. 4.11) are due to various cytochromes, the most distinct band is found at 550 nm and indicates the presence of the most abundant and widely occurring cytochrome, cytochrome *c*. These substances may also be identified in tissue extracts by their reaction with Nadis' reagent (Appendix, p. 254), which turns a pinkish-purple if cytochromes are present. The Cytochromes *a*, *b* and *c* act readily as carriers, as described above, in accepting electrons from dehydrogenase enzymes so that cytochrome oxidase, the terminal oxidase, may become oxidized by atmospheric oxygen.

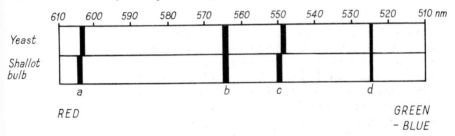

Fig. 4.11. Absorption spectra of the cytochromes. Each cytochrome has α, β, and γ bands; *a*, *b*, and *c* are the α-bands of cytochromes *a*, *b*, *c*, the band of cytochrome *c* being much the strongest. β bands appear towards the blue end of the spectrum about 525 nm at *d*; γ bands are found around 430 nm.

4.7   Oxidative phosphorylation

The glycolysis sequence includes two steps in which high-energy phosphate (ATP) is synthesized from ADP by non-oxidative steps; in the aerobic system a further 36 molecules of ATP are synthesized for each molecule of hexose respired. The problem is, how is this vital synthesis carried out? The answer is still by no means certain, but it is likely that a large amount of free energy (33 kJ/mole) is made available at each of two or three steps in each electron transfer chain. This energy is sufficient to allow for the formation of a high-energy phosphate bond. As the electrons flow from one substance to the next in the chain, the energy is transferred to the phosphorylation of ADP. This table summarizes the ATP formation in the breakdown of 2-oxopropanoic acid by the Krebs' Cycle:

Reaction	Molecules ATP synthesized
1. Oxidative decarboxylation of 2-oxopropanoic acid	3
2. 1-hydroxypropane-1,2,3-tricarboxylic acid → 1-oxopropane-1,2,3-tricarboxylic acid	3
3. Oxidative decarboxylation of 1-oxobutanedioic acid	4
4. Butanedioic acid → *trans*-butanedioic acid	2
5. 2-hydroxybutanedioic acid → 2-oxobutanedioic acid	3
Total	15

As two molecules of acid are oxidized for each molecule of hexose utilized, this gives a total of thirty molecules of ATP produced through the aerobic stages of respiration. The full total is thirty-eight molecules of ATP, two being produced during glycolysis and the remaining six being derived from the *oxidation* of the reduced coenzyme formed in the conversion of triose into phosphoglyceric acid. In effect, thirty-six molecules of ATP are synthesized by the aerobic steps.

The high-energy phosphate system has evolved in living things as a device for moving and using energy. Unfortunately these molecules are far too reactive to act as substantial stores of energy. The whole process of respiration has developed as a mechanism to convert the stored energy of carbohydrates, fats and other organic materials into ATP. This in turn provides the energy for growth, synthesis and movement that is the very essence of metabolism.

4.8   The site of respiration in the cell

Recently attention has been focused on the properties and organization of the various minute organelles within the cell. Phase-contrast films of cell-division usually show a very considerable activity of the small cigarshaped bodies called

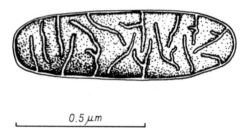

0.5 μm

Fig. 4.12. The mitochondrion.

mitochondria near the spindle of the dividing cell. These mitochondria (see fig. 4.12) show a very active jostling movement—rather more than would be expected to occur through Brownian movement alone—and it is possible that they are congregating where considerable energy may be needed in the process of cell-division. That they have an important part in cell activities, as the site of many respiratory enzymes and also of the high-energy phosphate synthesis system, has been shown by two main lines of research. In the first, bulk extracts of the mitochondria have been made and their properties examined; in the second, very thin sections have been photographed in the electron microscope and the structure of the mitochondria examined in detail.

The existence of the mitochondria has been known for many years, as in addition to phase-contrast studies, they may also be stained, though rather unevenly, by Janus Green B. Such staining has usually revealed the mitochondrion to be a minute elongated organelle about 0.7–1 μm in length. In spite of their small size, mitochondria can be extracted in bulk by fractional centrifuging. The first job is to extract the mitochondria as far as possible intact and undamaged. This can be done by careful grinding of the cells of the material using an ice-cold mixture of sucrose and phosphate buffer. This solution should have a slightly higher osmotic pressure than that of the original cell sap so as to

Fig. 4.13. Mitochondria may take a number of forms; in young cells they may have rather inflated cristae (C) as in fig. 4.13A. In other cells the mitochondria may themselves be in a state of reproduction by budding (fig. 4.13B) which shows the mitochondria with more typical cristae. Electron microscope photographs tend to show the mitochondria as round or elongated bodies, but evidence is beginning to accumulate that, at least in some instances they are much longer (fig. 4.13C). A young chloroplast (CH), endoplasmic reticulum (ER), ribosomes (R) and parts of the cell wall (CW) may also be seen in these photographs.

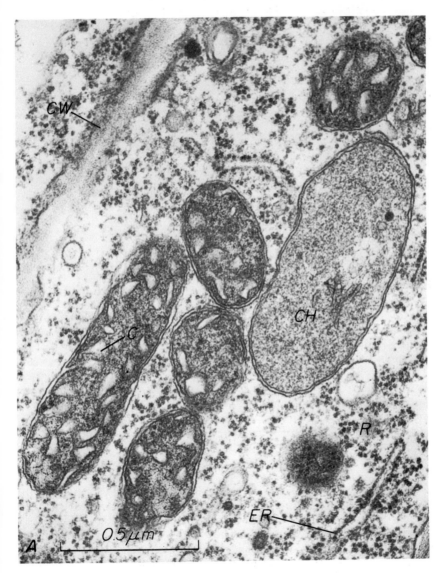

Fig. 4.13A. From the young leaf of maize. (Courtesy Dr Jean Whatley)

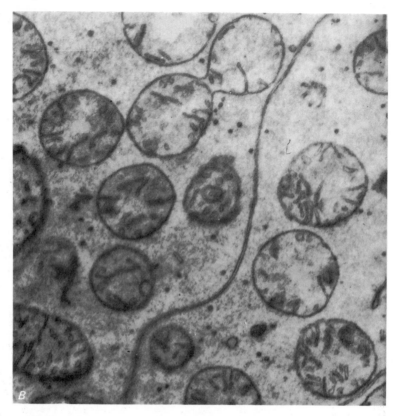

Fig. 4.13B. From the root tip of barley. (Courtesy B. E. Juniper)

prevent osmotic destruction of the organelles. The mitochondria are thrown down between 10 000 and 20 000 ×g, and after resuspension in buffer and washing by more gentle centrifuging are ready for examination.

Chemical analysis has shown them to be largely protein (30–40 per cent) and fat (25–38 per cent), the remainder being composed mostly of ribonucleic acid (RNA); spectroscopic analysis has shown also that the mitochondria are rich in the cytochromes. Addition of 2-oxopropanoic acid to the preparation of mitochondria results in its oxidation, while addition of radioactively labelled butanedioic acid, with chromatographic and autoradiographic analysis, shows the presence of labelled trans-butanedioic acid and 2-hydroxybutanedioic acid. Therefore the mitochondria probably have most of the enzyme systems of the Krebs' Cycle, though it is of course possible that other parts of the cell also produce similar enzymes. A most interesting and important feature of the mitochondria is that ADP and free phosphate ions seem to be necessary for their efficiency in oxidation; they are in fact the site where the ATP is synthesized.

Electron-microscope photographs of ultra-thin sections of mitochondria have shown them to be composed of a thin outer membrane and a more complex inner system composed of channels called *cristae* running into the inner part of the organelle. However, the structure of mitochondria is variable; for instance, see

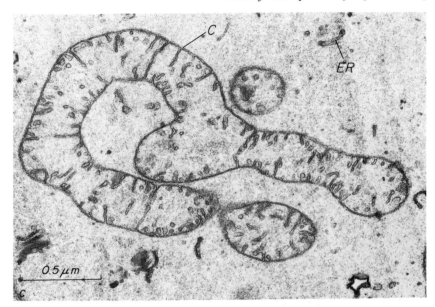

Fig. 4.13C. From the root cap core cells of maize. (Courtesy of A. Bennell)

fig. 4.13. Some idea of how and where these vital respiratory steps take place has been gained from autoradiographs taken after feeding with radioactive intermediates. Another technique, using tellurium oxide or tellurite has also proved useful. The oxide is reduced to the metal (which is dense to electrons and appears black in electron-microscope photographs) by the action of the reduced butanedioic dehydrogenase formed in the butanedioic acid–*trans*-butanedioic acid reaction in the Krebs' Cycle. Accordingly, wherever the butanedioic acid reaction takes place, black specks of tellurium metal may be expected, and indeed these are found lining the cristae of the mitochondrion. It seems likely that these indentations may be the site where the vital steps of respiration, resulting in the formation of ATP, take place. As electron transfer systems are vital to these steps, it is thought that the arrangement of the cristae is such as to allow ready exchange of electrons between enzyme and substrate by providing a suitable space for electron transfer to take place.

4.9 Summary and importance of respiration

Respiration consists in its early stages of a reversal of photosynthesis; the hexose sugar (fructose-diphosphate) is broken down through three carbon sugars to phosphoglyceric acid, and this is eventually converted into 2-oxopropanoic acid. Little energy in the form of ATP is made available in this anaerobic process of glycolysis. The process is rather a rearrangement of the molecules concerned into a position from which efficient breakdown resulting in the formation of high-energy phosphate can take place. In the anaerobic system two molecules of ATP are synthesized for each molecule of hexose originally utilized, while in the aerobic Krebs' Cycle system 36 molecules of ATP are synthesized for each molecule of hexose. This illustrates the much greater efficiency, in terms of ATP synthesis, of the aerobic system. The actual efficiency of energy yield in aerobic respiration is surprisingly high, being about 67 per cent.

The energy relationships of the plant can be summarized into four main phases (see fig. 4.14). In the first place the sun's energy is captured by the photosynthetic pigments. Secondly, the energy is stored in *stable* complex organic substances. Thirdly, these substances are modified into *labile* forms (e.g. 2-oxopropanoic acid) which can be broken down to release energy, which is transferred to form

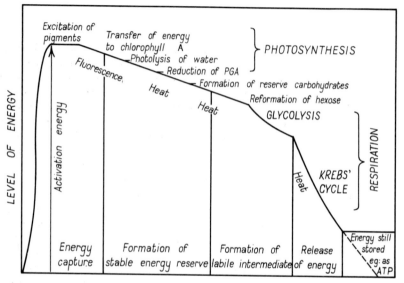

Fig. 4.14. The energy relationships of the plant. The captions under the curve indicate energy loss that is necessitated by the laws of thermodynamics.

ATP. At each step some energy is lost—a necessity in terms of the second law of thermodynamics—but the net effect is that the ATP is formed which is capable of activating further energy-requiring processes.

The fascinating part of respiration undoubtedly concerns the formation of this substance, and interest in respiration is now directed towards unravelling the mechanism through which this occurs.

Further reading on Respiration

BRYANT, C. (1971). *The Biology of Respiration.* Institute of Biology's Studies in Biology no. 28. London, Edward Arnold.
JAMES, W. O. (1971). *Cell Respiration.* London, English Universities Press.
TRIBE, M. A. and WHITTAKER, P. A. (1972). *Chloroplasts and Mitochondria.* Institute of Biology's Studies in Biology no. 31. London, Edward Arnold.

5

Mineral Nutrition

5.1 Minerals and the soil

One of the main factors influencing plant distribution is the type of soil. In the British Isles there are a number of more or less clearly defined soil types. One of the simplest of these is the *rendzina*. This is found on chalk and limestone rocks, where the humus layer is thin and lies directly on the rock substrate. Such a soil is kept in this state by the rapid-draining, porous nature of the rock, which keeps the humus layers dry and aerated and so promotes the oxidation of the decaying material. A chalk soil, being so close to the parent rock, is rich in minerals, particularly calcium and magnesium ions, but on the other hand, the high pH may render other minerals insoluble. For instance, the important elements iron and manganese are relatively unavailable to plants.

The group of plants that are adapted to living under such soil conditions are called *calcicoles*. Typical plants in this group include the upright brome (*Bromus erectus*), small scabious (*Scabiosa columbaria*), salad burnet (*Poterium sanguisorba*) and squinancywort (*Asperula cynanchica*). These are well adapted to life on calcareous soils, as they have low demands for iron and manganese. Many calcicoles can, however, live in acid soils provided aluminium is absent. This element is more readily available on soils of low pH and if absorbed by calcicoles may prevent their growing.

A complete contrast is the *podsol* soil found in sand and gravelly areas. Here the soil is open and porous, but there are relatively few minerals present and these are easily washed or leached out of the soil by the rain. High humidity may favour the accumulation of undecayed humus, which is acid and aggravates the leaching effect. On these acid soils calcium and magnesium have usually been dissolved away, but iron and manganese are freely available, the latter being in the readily absorbed bivalent state. Sometimes too much manganese may be available, which may have a toxic effect. Plants which are adapted to life on acid soils of this sort are called *calcifuges*. Ling (*Calluna vulgaris*), heath bedstraw (*Galium saxatile*) and the grass *Holcus mollis* are in this class. These plants have high demands for potassium and phosphate ions, which are less readily absorbed in the presence of calcium ions, which tend to antagonize their uptake. In addition, they can tolerate the aluminium level of the soil and have fair demands for the readily available iron and manganese.

The most suitable soil for agriculture lies half-way between these two soil types. This is the *brown earth*. This type of soil is usually found on clays or rich sands and has a pH near neutrality, about 6.5. Here there is a fair balance between the mineral content, the accumulation of humus and the water content, so that a reasonable quantity of dissolved ions are available to plants. Such soils support the rich vegetation of the natural oak woods, but nowadays they are mostly exploited as arable land.

These general observations on soil minerals emphasize several important features of mineral nutrition. Most plants are specially adapted to more or less specific soil types. These soils possess a *suitable concentration* of the minerals which that species particularly needs; too much may have toxic effects, too little may result in poor growth. A second point is that the correct *ion balance* must exist in the soil. Lack of one element may result in another element, normally useful to the plant, acting in a toxic manner. This is illustrated by the fact that calcicoles may be unable to grow on calcium-deficient soils because too much potassium is absorbed. Similarly, the water cultures described below will grow plants well only if there is the correct physiological balance of nutrients.

Antagonism may occur between two similar ions. Such *ion competition* may take place if both ions are absorbed into the root cells by a similar carrier system; both ions appear to compete for the same site on the carrier. For instance, the presence of potassium antagonizes magnesium uptake, while uptake of rubidium or calcium affects potassium uptake, although, perhaps surprisingly, sodium does not interfere with potassium and must presumably be taken up through a different carrier system.

The study of the soil and the minerals it contains is a vital part of the investigation of plant mineral nutrition; on the other hand, it is equally important to know which minerals are taken up and how they are utilized in plant metabolism before conclusions can be drawn about the economic and effective use of fertilizers and soil improvement techniques.

5.2   The mineral requirement

In addition to the elements carbon, hydrogen and oxygen that may be absorbed as water, carbon dioxide or oxygen, and which together make up a large part of the weight of a plant, there are a large number of mineral elements which are as necessary for plant growth. This necessity for various mineral elements has been investigated by the use of water cultures. A complete nutrient medium can be made up containing all the minerals that are thought to be necessary, and then separate media are made up omitting one nutrient mineral in each case (see Appendix, p. 235. The effect of different culture solutions on the size of young barley plants is illustrated in fig. 5.1. Many differences in addition to those of size are found in this type of experiment. Calcium, potassium, nitrogen and iron are particularly important in the growth of barley. The presence and amounts of the various minerals actually metabolized can be found by drying, washing and subsequent analysis. Fig. 5.2 illustrates the amounts of these minerals in young plants of winter wheat; again nitrogen and potassium are particularly important.

Experiments on the mineral requirements of lower organisms such as fungi and bacteria are similar in principle and are particularly useful, as these organisms have a quick growth (or reproductive) rate. Sterile agar is used as the medium for growth, and the organisms are grown in flat Petri dishes. (The preparation of the nutrient media is described in the Appendix on p. 236) If the fungus *Mucor* is grown on a series of different nutrient media the effect on the

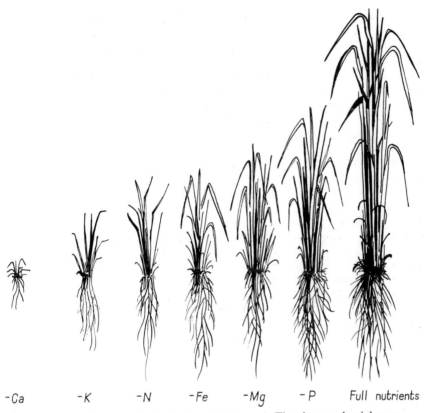

−Ca   −K   −N   −Fe   −Mg   −P   Full nutrients

Fig. 5.1. The nutrients required by barley (*Hordeum* sp.). The plant on the right was provided with all the minerals essential for plant growth. The remainder had similar nutrients except that the mineral indicated was omitted from the culture solution.

growth rate is very marked; phosphate and ammonium ions are particularly useful for this species (see fig. 5.3).

Techniques of this sort have shown that there are six essential *major elements* which are required in fairly large quantities (see p. 96). These are nitrogen, potassium, calcium, phosphorus, magnesium and sulphur. Certain *minor* or *trace elements* may also be classified as essential in that most plants will not grow satisfactorily in their absence. These are iron, chlorine, copper, manganese, zinc, molybdenum, boron, cobalt, sodium, vanadium, gallium, silicon and iodine. These elements are only required in very small quantities and in some cases the requirement is so small that it is extremely hard to show it because of the difficulty of providing conditions in which the relevant element is absent. Some other elements (e.g. aluminium, rubidium, selenium) may be beneficial to growth in certain plants but have not been shown to be essential. There is sometimes difficulty in classifying an element as 'essential' because in a number of cases, such as sodium and silicon, the element may be needed for some plants but not others. (See fig. 5.4 and p. 96.)

Sodium is frequently absorbed by plants and, while probably not essential in most cases, is sometimes regarded as beneficial for plant growth. On the other hand, in the salt marsh plants, the halophytes (see p. 34), it is taken up in large quantities and is probably an essential requirement. Silicon is absorbed by

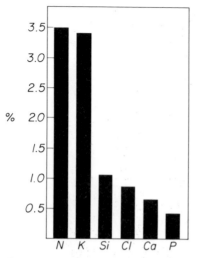

Fig. 5.2. The mineral content of winter wheat plants. Relative amounts of various minerals in young winter wheat plants sampled in April, expressed as per cent of total dry matter.

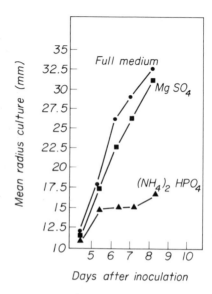

Days after inoculation

Fig. 5.3. Growth of *Mucor* with the mineral nutrients indicated omitted from the culture medium. (Class result.)

grasses in some quantity, but most of these seem to be able to live healthily without it. Aluminium is also frequently absorbed, sometimes with toxic results, causing upsets to the cell-division system, at other times causing no apparent damage.

The table on page 96–97 summarizes the importance of the various minerals utilized by plants.

Increasing concern about environmental pollution has led to more attention being given to the reclamation of land contaminated by industrial wastes. Studies have been carried out on the accumulation by some plants of unusually high levels of elements that are normally toxic at high concentrations. Examples of these are copper, zinc, lead, chromium and nickel. Part of the interest in this subject derives from the fact that in some plants, especially grasses, races have evolved which are tolerant of such elements. This evolution has clearly been very rapid, because many of the sites where the plants are found are of recent origin. The possible usefulness of such plants in reclamation work is clear, though care must be taken that such plants are not used extensively as animal foodstuffs.

5.3   Uptake of minerals

There are four main stages in mineral uptake. First, that in which mineral ions diffuse from the soil solution to the root hairs and into the intercellular spaces of the root apex. In the second stage ions may become adsorbed on to the cell walls,

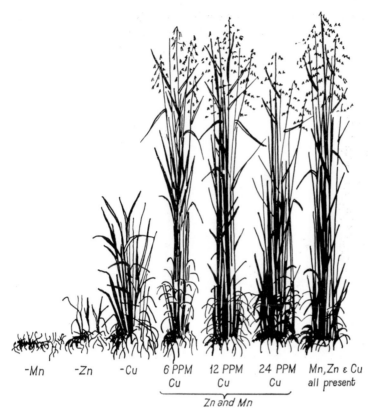

-Mn   -Zn   -Cu   6 PPM   12 PPM   24 PPM   Mn,Zn ε Cu
                            Cu   Cu   Cu   all present

Zn and Mn

Fig. 5.4. Some minor nutrients essential for plant growth (oats). The plant on the right was provided with all the minerals essential for plant growth. The three on the left had similar nutrients but lacked the one indicated. The remaining plants had a full medium, but with different concentrations of copper. Note that there is an *optimum concentration* for copper; too much has a mildly detrimental effect.

and in the third stage may diffuse through the cell wall and outer cytoplasmic membrane (plasma membrane) into the cytoplasm. Finally, if the plant is somewhere starved of minerals these ions may accumulate in the vacuole of that cell, but, on the other hand, if the vacuole already contains a high concentration of ions, then they may be transferred to the cytoplasm of the next cell directly, probably through the cell–cell protoplasmic connections, the plasmodesmata (see fig. 6.9).

The uptake of minerals into both cytoplasm and vacuoles is probably best investigated in the first instance, by extending the culture solution experiments described above so that changes in external nutrient concentration are noted. It is found that the concentrations of many of the minerals are radically changed during the course of the experiment; fig. 5.5 illustrates this with barley. Potassium ions are entirely absorbed, nitrate is also strongly absorbed and other elements relatively less. Clearly some minerals are taken up more readily than others; the problem is, how are they absorbed and why are some selectively absorbed in preference to others?

A Summary of Minerals used by Plants

Element	Form in which absorbed	Quantity utilized (approx.) as per cent of dry wt of plant	Principal Functions	Effect of deficiency	Ecological and agricultural notes	Fertilizers
MAJOR ELEMENTS						
1 Nitrogen	NO_3^- (or NH_4^+)	3–5	Amino-acids, proteins and nucleotides	Chlorosis, small-sized plants	Frequently deficient; organic manuring and addition of nitrogenous fertilizers often necessary	Ammonium sulphate, sodium nitrate, nitro-chalk (ammonium nitrate and fine chalk), compost, farmyard manure
2 Potassium	K^+	3–4	Enzyme, amino-acid and protein synthesis. Cell membranes. Stomatal opening and closing, osmotic relations of cells	Leaves have yellow edges; premature death	Plants need more potassium after heavy manuring with nitrogen and phosphorus. Most available on acid soils, though it may be leached out	Potassium sulphate
3 Calcium	Ca^{2+}	0.7	Calcium pectate of cell-walls. Development of stem and root apices. Enzyme activation	Stunting of the root and stem	Little present in acid soils. Has important effect on soil by assisting flocculation of clay particles	Nitro-chalk, calcium phosphate, basic slag, superphosphate of lime [$Ca(H_2PO_4)_2 + CaSo_4$]
4 Phosphorus	$H_2PO_4^-$ (ortho-phosphate)	0.4	Involved in energy transfer (ATP and ADP). Nucleic acids. Phosphorylation of sugars. Coenzymes	Small-sized plants; leaves a dull, dark green	Frequently deficient. Little is available over pH 7	Superphosphate of lime, calcium phosphate, basic slag
5 Magnesium	Mg^{2+}	Small quantity	Part of the chlorophyll molecule. Activator of some of the enzymes in phosphate metabolism	Chlorosis; older leaves turn yellow, their veins remain green	Often deficient on acid soils	Magnesium sulphate, basic slag
6 Sulphur	SO_4^{2-}	Small quantity	Proteins which contain thiol (–SH) groups. Active groups in enzymes and coenzymes	General chlorosis	Seldom deficient in Great Britain due to sulphuric acid contained in atmospheric pollution	—
MINOR ELEMENTS						
7 Iron	Fe^{2+}	Small quantity	Chlorophyll synthesis. Cytochromes and some enzymes	Chlorosis, young leaves turn yellow-white, their veins remain green	Much less available on calcareous soils, being in the form of insoluble ferric hydroxide	Chelated iron (in which the iron is bonded to an organic molecule), e.g. sequestrene 138 Fe
8 Chlorine	Cl^-	0.8	Osmosis and anion/cation balance; probably essential in photosynthesis in the reac-	Effects slight	—	—

		0.0001 per cent	systems. Photosynthesis			
10 Manganese	Mn^{2+}	,,	Activator of some enzymes (e.g. carboxylases). Photosynthesis	Chlorosis and grey-specks on leaves	Manganese is in the divalent state in acid soils and is readily available even to a toxic level. At high pH manganese is in the trivalent state, which plants cannot utilize	—
11 Zinc	Zn^{2+}	,,	,,	Leaf malformation	More often deficient on acid soils due to adsorption on to colloidal complexes in soil	—
12 Molybdenum	Mo^{3+} or $^{4+}$	,,	Nitrogen fixation and metabolism, enzyme nitrate reductase	Size of plants slightly reduced	If molybdenum is deficient plants may survive if nitrogen is supplied as NH_4^+, but not if supplied as NO_3^-	—
13 Boron	BO_3^{3-} or $B_4O_7^{-}$ (borate or tetra-borate)	,,	Influences Ca^{2+} uptake and utilization. Differentiation and pollen germination. Sugar transport	Brown heart disease	Easily leached from soils, particularly those of low pH	—
14 Cobalt	Co^{2+}	,,	Various roles in symbiotic nitrogen-fixing plants	—		—
15 Sodium	Na^+	Trace	Osmotic and anion/cation balance, probably not essential to most plants, though essential to some	Effects slight		—
16 Vanadium	V^{2+}	,,	Not known			—
17 Gallium	Ga^{3+}	,,	Not known			—
18 Silicon	$H_2SiO_4^{2-}$	1.0 (grasses)	Straw formation (calcium silicates). Not essential to most plants, though essential to some (e.g. grasses)	Slight decrease in weight		—
19 Iodine	I^-	Trace	Not known			—
NON-ESSENTIAL ELEMENTS						
20 Aluminium	Al^{3+}	,,	Not essential. May cause upset to cell division system		More available in acid soils; it may prevent the growth of calcicoles	—
21 Fluorine, nickel	F^- Ni^{2+}	,,	Not known, but possibly essential in some cases			—

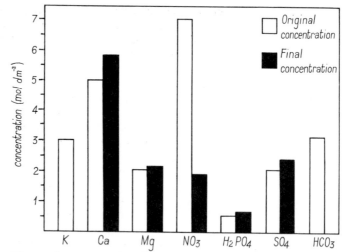

Fig. 5.5. Changes in concentration of nutrient solution due to uptake of water and ions by barley plants. (After Brierley (1958). An Approach to the Teaching of Salt Uptake in Plants. *School Science Review*, No. 138, p. 254.)

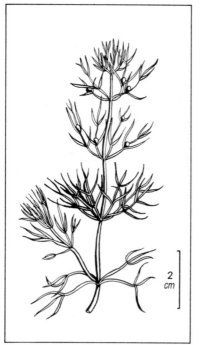

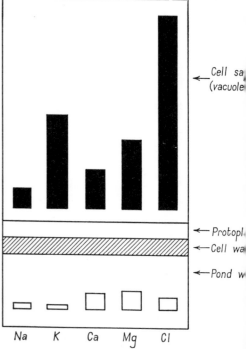

Fig. 5.6. *Nitella.* A coenocytic alga useful in investigation of the concentration of ions in the vacuole.

Fig. 5.7. Relative concentrations of different ions in pond water and the vacuolar sap of *Nitella.* (After Brierley (1958). An Approach to the Teaching of Salt Uptake in Plants. *School Science Review*, No. 138, p. 25

The next step is to determine the internal and external concentrations of ions to find out the magnitude of any concentration difference, as it may be difficult for ions to diffuse into a vacuole which already contains more dissolved material than the external solution. Work with the large fresh-water algae *Valonia* and *Nitella* (see fig. 5.6) has proved to be particularly useful, as their cell sap is easily extracted and the internal concentration can be compared with that of the pond water in which they were growing (see fig. 5.7). In this species it is clear that sodium, potassium, calcium, magnesium and chloride ions are all at a much higher concentration in the vacuole than in the pond water, but again all the ions are not absorbed to the same extent. These results emphasize two important features of mineral uptake: first that the cytoplasmic membrane has some system by which ions are selectively absorbed, and secondly, that if this diffusion against a concentration gradient is to take place, then some energy will be required to drive the ions into the stronger internal solution of the vacuole. Similar results have been obtained with higher plants using the sap collected from cut surfaces and by rapid freezing techniques, the concentration differences being even more marked.

It used to be thought that the principal sources of this energy came from the thermal energy of the ions themselves and that once the ions reached the vacuole they were adsorbed and in some way taken out of free solution, thus lowering the internal concentraton and allowing more ions to enter. However, experiments on the conductivity of the cell sap of algal coenocytes have indicated that the ions are in free solution and in no way adsorbed. Another indication that the ions are in free solution is that the osmotic potential of the sap falls after a period of salt uptake.

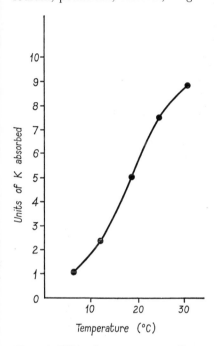

Fig. 5.8. Effect of temperature on the absorption of potassium by excised barley roots. (After Brierley (1958). An Approach to the Teaching of Salt Uptake in Plants. *School Science Review*, No. 138, p. 254.)

Mineral uptake is a characteristic of actively metabolizing tissues; this suggests that the energy required to drive the ions into the vacuole comes from some active metabolic process such as respiration. This suggestion is supported by an experiment in which the absorption of potassium by young barley root tips is related to temperature (see fig. 5.8). This type of curve is typical for enzyme-controlled processes such as respiration, in which the rate of the reaction about doubles itself for a rise in temperature of $10°C$.; in other words, the Q_{10} is near 2. The rate of a physical process such as the diffusion of ions does not rise nearly so fast, the Q_{10} being about 1.3.

If respiration is involved, then the effect of oxygen on the rate of uptake should be marked. A simple experiment involves the uptake of bromide ions by young barley root tips in air and in nitrogen (for full details see Appendix, p. 237). At the same time the temperature effect can be examined. Absorption of bromide in nitrogen is only about 10 per cent of that in air at room temperature; full results are given in the table below:

	Temperature (°C)	Uptake (expressed as a percentage of that in air)
Air	17	100
Nitrogen	17	10
Air	3	28

Table to show the absorption of bromide by barley root tips under different conditions (Class results.)

This experiment shows that aerobic respiration is a necessity for any reasonable uptake of ions such as bromide.

Further evidence of the close relationship between respiration and salt uptake comes from observations on the effect of adding a salt to a respiring tissue and estimating any changes in the respiratory rate. Very often a stimulation of the respiratory rate follows; this is known as the *salt respiration effect*. However, not all salts produce the effect; this is probably due to their specific effects on the absorbing system in the cytoplasm. Part A of the graph (fig. 5.9) illustrates the rate of respiration of the yeast-like fungus, *Torilopsis utilis*, when a salt is added. It must also be emphasized that the concentration of salt that has to be added to produce the salt respiration effect is much higher than would exist in the soil solution. In this species sodium has no effect, but ammonium phosphate causes a considerable stimulation. That the ammonium phosphate is absorbed by the fungus and at the same time removed from the solution is shown by separate estimation (see Appendix, p. 238), the results of which are shown in parts C and B of the graph, fig. 5.9.

The connection of salt uptake with respiration can be further investigated by the use of inhibitors. For instance, 10^{-3}M potassium cyanide, which inhibits the terminal oxidase systems with copper and iron activator groups, also has a considerable effect on the rate of salt uptake.

Evidence about the mechanism of ion uptake has come from the use of radioactive ions whose movements from the outside medium into the cell can be traced. This technique is particularly useful since intact plants can be used. Some ions may move across the plasma membrane passively in response to a diffusion or electrical potential across this membrane. Radioactive rubidium, a metal which appears to be taken up in much the same way as potassium, is a case in point, though its movement across the tonoplast into the vacuole is not a passive process. Other ions cross the plasma membrane against a gradient of diffusion or electrical potential, and this process clearly requires the expenditure of energy as does the movement of radioactive rubidium in the previous example. The energy responsible for driving this movement must come ultimately from the photosynthetic and respiratory systems, though the precise mechanisms by

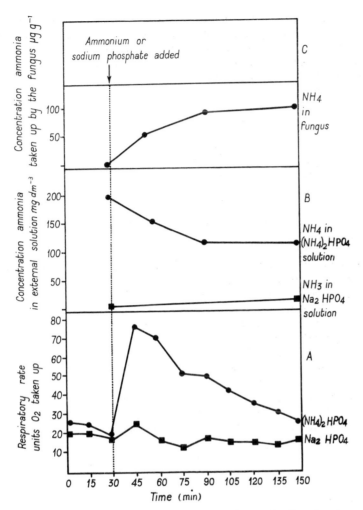

Fig. 5.9. Salt respiration and the absorption of ammonium and sodium ions (see p. 238)

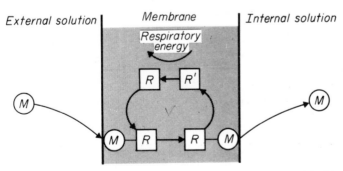

Fig. 5.10. *The carrier hypothesis.* M is the penetrating ion, R the carrier, M-R the mobile complex and R' a mobile carrier precursor. (Redrawn after Sutcliffe and Baker, 1974.)

which the process of transfer across the membrane occurs is still not known. One hypothesis is that there exist in the membrane 'carrier' molecules, which are probably proteins; these selectively bind on to an ion on one side of the membrane, move it to the other side and then release it. This process is illustrated in fig. 5.10.

One advantage of this hypothesis is that it is possible to involve different carrier molecules for different ions and thus explain the differing behaviour of membranes relative to such ions. Energy could be used at various stages of the process, for example in the regeneration of the carrier R from the precursor R′. The carrier hypothesis has many attractive features but is not universally supported and we still do not really understand the process of ion movement across membranes against electro-chemical gradients. It seems extremely likely that more than one mechanism may be operating (see also p. 147).

5.4   **Mineral transport** (see also p. 31)

Once the ions have been absorbed from the soil solution they may pass into the intercellular spaces, cell walls, cytoplasm or vacuoles of the cells behind the root apex. How then are they transported to areas where they are required? The first phase of this transport must end at the xylem. Although some movement of ions takes place through the apoplast (see p. 29), much of the transport may occur through the living system of the *symplast*. If the plant is well supplied with minerals this takes place from the cytoplasm of one cell direct to the cytoplasm of another, through the plasmodesmata, without many of the ions being

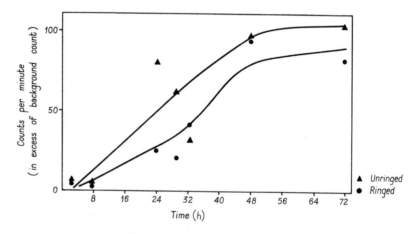

Fig. 5.11. Accumulation of ^{32}P by ringed and unringed plants of *Skimmia*. (Class results.)

temporarily accumulated in the intervening vacuoles. Most of the ions are transported in the form in which they are absorbed, but nitrogen is transported primarily as amides and amino-acids; how they are moved into the xylem is not known.

Once the ions reach the xylem they are drawn upwards in the xylem stream. That it is the xylem and not the phloem that is principally important for this upwards movement can be shown by ringing experiments, in which the phloem elements are removed. This can be demonstrated most effectively by placing two similar plants of *Skimmia* or other woody plants in a solution containing radioactive phosphorous, ^{32}P, after one plant has had its stem ringed. The counts per minute near the apex of each plant are measured with a Geiger counter until sufficient tracer has accumulated (see fig. 5.11). After about three days both plants are removed from the solution and their aerial parts placed for about two weeks against a photographic plate. After development it is seen that both autoradiographs show tracer in their aerial parts.

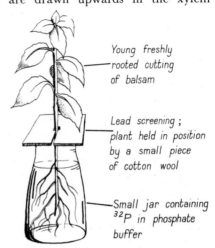

Young freshly rooted cutting of balsam

Lead screening; plant held in position by a small piece of cotton wool

Small jar containing ^{32}P in phosphate buffer

Fig. 5.12. Apparatus used for investigating the absorption of ^{32}P by young plants of balsam (*Impatiens sultani*).

However, phloem transport also takes place, especially in tissues where the transpiration stream is low, such as the apical meristems; these are also tissues in which metabolism of the minerals is vitally important. That lateral diffusion from the xylem to the phloem normally takes place is well shown by another tracer experiment involving the use of radioactive phosphorus (for full details of the use of radioactive tracers see Appendix, p. 239). A small shrub such as a willow with three shoots is allowed to grow in a solution containing radioactive phosphorus. One shoot is left intact as a control; in the second the xylem and phloem are separated by a piece of thick waxed paper and then bound together; in the third the xylem and phloem are separated and then rejoined, without the waxed paper, as another control. After about twenty-four hours the treated segments are sampled, sectioned lengthways and the flat surfaces placed on photographic plates. The resulting autoradiographs show that in the first and last segments the tracer is uniformly distributed in both xylem and phloem, but in the second very little has passed into the phloem. This indicates that although the xylem is the main path of transport of ions, lateral diffusion into the phloem does take place to a considerable extent.

The leaf petiole is a structure which has also been investigated with reference to the transport of mineral and organic molecules. Hot wax and heat jackets surrounding the petioles, causing coagulation of the protoplasm of the living phloem elements, have been found to prevent the downwards transport of organic molecules from the leaf but to leave the uptake of minerals into the leaf unaffected. This has been shown using radio-active phosphorus, ^{32}P, its presence in the leaf being identified autoradiographically or with a counter.

Although it looks as though the xylem is usually the most important route for the upwards transport of mineral ions, there is some evidence, obtained from ringing experiments, that the phloem is of comparatively greater importance in this role in woody plants. Ringing experiments on trees have been shown to have a marked effect on the upward transport of amides.

Once the mineral ions reach the leaf and aerial parts of the plant they diffuse out of the finest xylem elements and are finally taken into the cells, where they are required for their various metabolic roles. Studies of autoradiographs of

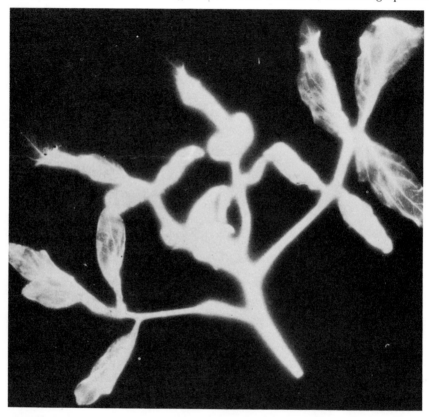

Fig. 5.13. Autoradiograph showing ^{32}P accumulation by tomato. An intact, rooted seedling was placed for two days in a solution containing six microcuries of ^{32}P. The roots were then removed and the autoradiograph obtained by placing a photographic plate over the flattened plant. Tracer has accumulated throughout the leaves, though most strongly in the veins. (Positive print from autoradiograph.)

plants that have been growing in various tracer elements show to some extent how the ions are distributed, and can be a great help in understanding the mineral nutrition of plants. Balsam and tomato provide useful material (see figs. 5.12 and 5.13), as uptake is particularly rapid, phosphate being distributed throughout the leaves, after an initial accumulation in the growing apices.

5.5   Metabolic utilization of mineral ions

Once the minerals are in the xylem they are transported in the transpiration stream to the stems and leaves. In the young stem apices rapid growth is taking place and active transport of the ions may be carried out so that these minerals can be incorporated into the various substances and structures being laid down. Nitrogen, arriving in various forms, is metabolized into amino-acids, proteins and other compounds. The enzyme nitrate reductase (which requires molybdenum as its activator group) reduces nitrate to ammonium ions. These cause the amination of the Krebs' Cycle intermediates, such as 1-oxobutanedioic acid to form 2-aminopentanedioic acid.

$$NO_3^- \longrightarrow NH_4^+$$

nitrate
reductase
(Mo ions needed)

$+$

$$
\begin{array}{l}
COOH \\
| \\
CH_2 \\
| \\
CH_2 \\
| \\
CO \\
| \\
COOH
\end{array}
$$
1-oxobutanedioic acid

$$
\begin{array}{l}
COOH \\
| \\
CH_2 \\
| \\
CH_2 \\
| \\
CH . NH_2 \\
| \\
COOH
\end{array}
$$
2-aminopentanedioic acid

Phosphate ions may exist free in the cytoplasm for any length of time and are continually being used and reformed through the various phosphorylation reactions. Considerable quantities of phosphate are continually required for the various phtosynthetic and respiratory intermediates as well as for the formation of the high-energy phosphates (ATP and ADP). It is possible to distinguish those ions which have a structural or metabolic role: particularly nitrate, phosphate, calcium and sulphate. Other ions, such as magnesium, iron and copper, are vital as parts of enzymes or catalysts and are called *catalytic ions*. The division of mineral ions into these groups bears a relationship to the qantities of minerals that are required. Those with a catalytic role are needed in only small quantities compared with those concerned in the production of structural or metabolic substances.

5.6 The nitrogen cycle

It is probably a mistake to suggest that one element is more important in plant metabolism than another, as varying quantities of all the minerals mentioned above are essential to plant growth. However, the turnover of nitrogen in plants and in the soil is rightly regarded as one of the most important aspects of mineral nutrition.

Relatively few plants are able to utilize atmospheric nitrogen directly (see nitrogen-fixing organisms, below), but are dependent on nitrate for most of their nitrogen requirements, although ammonia is also absorbed. Many soils contain some inorganic nitrogen compounds derived from fertilizers, from atmospheric fixation due to electrical storms and from the decay of plant and animal materials. Saprophytic bacteria, such as *Bacillus megatherium*, and various species of fungi, are mainly responsible for the final breakdown of dead organic material resulting in the release of ammonia into the soil solution. The activities of saprophytic fungi are particularly noticeable in woodland. In beechwoods in the autumn there are to be found a wide selection of fruiting-bodies of litter-decomposing species, such as the yellow-staining mushroom (*Agaricus xanthoderma*), the small purple-pink *Mycena pura* and the brown *Marasmius peronatus*. Wood-decomposing species that are common include the bracket fungi, *Trametes versicolor* and *Trametes gibbosa*.

Although large quantities of ammonia are toxic to plant life, several species of bacteria that are common in soils are able to oxidize ammonia. The chemosynthetic nitrifying bacteria *Nitrosomonas* and *Nitrococcus* oxidize the ammonia to nitrite to release energy which they utilize for carbohydrate synthesis:

$$2NH_3^+ + 3O_2 \rightarrow 2NO_2^- + 2H^+ + 2H_2O$$

Another common chemosynthetic bacterium, *Nitrobacter*, oxidizes the nitrite to form nitrate, again with the release of energy which is utilized by the bacteria:

$$2NO_2^- + O_2 \rightarrow 2NO_3^-$$

The nitrate ions so formed are easily absorbed by plants. Although these equations are probably over-simplifications, they are useful in showing how chemosynthetic organisms may obtain their energy.

Another effect which is important in adding to the level of nitrogen compounds in the soil is the formation of nitric oxide and ammonia by high-tension discharges in the atmosphere; these substances dissolve easily in the rain. The nitrogen-fixing micro-organisms are, however, the chief means by which the level of fixed nitrogen in the soil is maintained. It is of interest that the fixation of atmospheric nitrogen by living things seems to be confined to micro-organisms; many examples have been listed of bacteria, algae and fungi being able to fix nitrogen, but so far there is no evidence that the higher plants, on their own, are able to carry out the process. On the other hand, lower organisms in a symbiotic association with other plants are frequently able to fix atmospheric nitrogen, even though both micro-organism and higher plant, when on their own, are usually unable to do so.

Unfortunately several other species of bacteria (e.g. *Micrococcus denitrificans*) are harmful, for they are able to reverse the processes of nitrogen fixation and nitrification. Fortunately they work effectively only in badly drained, oxygen-deficient soils.

The availability of fixed nitrogen in the soil is vitally important to most plants. The effects of growing leguminous plants in the ground in which barley is subsequently sown are shown in the table below and the turnover of nitrogen in living things is illustrated by the Nitrogen Cycle.

Legume	Legume nitrogen added to the soil in 1952 (kg ha^{-1})	Total nitrogen in barley plants in 1953 (kg ha^{-1})
Lucerne (alfafa) (Ranger)	12.4	7.2
Sweet clover (Hubam)	9.6	8.0
Ammonium nitrate (22 kg ha^{-1})	—	8.4
No treatment	—	5.6

The effect of legume green manure on the subsequent nitrogen yield of barley. (After Kroonje and Kehr 1956)

5.7 Nitrogen-fixing organisms

Much difficulty has been experienced in the past in determining the ability of organisms to fix nitrogen. There are several reasons for this; one is that some organisms may show an apparent growth without the synthesis of any new protoplasmic material, in other cases organisms have been able to obtain fixed nitrogen from the small amounts of the oxides of nitrogen existing in the atmosphere. In addition, contamination of cultures has sometimes caused confusion as another organism may be responsible for the fixation. Finally, difficulties have been experienced in the analysis of nitrogen actually fixed, and in the preparation of nitrogen-free culture media. A useful technique for the analysis of fixed nitrogen is the micro-Kjeldahl method, but recently useful results have been obtained with ^{15}N tracer nitrogen, which can be analysed using a mass-spectrometer. In recent years rapid advances have been made possible by the discovery that the enzyme systems responsible for nitrogen fixation will also convert ethyne to ethene, which can be easily detected in small quantities using gas chromatography techniques.

Fixation of nitrogen is a surprisingly frequent feature of free-living lower organisms. It is found in some species of saprophytic soil bacteria, photosynthetic bacteria, chemosynthetic bacteria and blue-green algae. Of the bacteria the three genera *Clostridium*, *Azotobacter* and *Beijerinckia* contain species that are nitrogen fixing. All these genera are widespread and common; *Clostridium* is particularly interesting, as it is an obligate anaerobe. The other two genera are aerobic, but *Azotobacter* tends to live in less-acid soils than *Beijerinckia*. Several photosynthetic bacteria are nitrogen fixing, but *Rhodospirillum rubrum* is of special interest, as fixation actually appears to be associated with the photosynthetic process. *Desulphovibrio* is an economically important chemosynthetic and obligately anaerobic organism that is also nitrogen fixing. About twenty species of blue-green algae are known to be nitrogen fixing. Many of these occur free in the soil and in ponds and streams, but they are also commonly found in symbiotic associations: *Nostoc* and *Anabaena* are well-known examples.

Fixation of nitrogen by symbiotic association is of particular biological interest, and includes the well-known examples of fixation in the root nodules of leguminous plants. Under normal conditions bacteria of the genus *Rhizobium*, which on their own are unable to fix atmospheric nitrogen, infect the root hairs of young leguminous plants and cause the formation of nodules. The bacteria stimulate meristematic activity in the outer cortex of the host and a nodule begins its development. Differentiation from the apex of the nodule (see fig. 5.14) results in the formation of several distinct zones. First, there is the uninfected apical meristem itself, then a zone of newly infected tissue followed by a large area of dense, active bacterial tissue and finally a zone of tissue disintegration. At the same time the apex forms a cortex, endodermis and vascular strand surrounding the central bacterial tissue and allowing for its protection and nourishment. The degree of infection is very varied (see fig. 5.14) and depends on

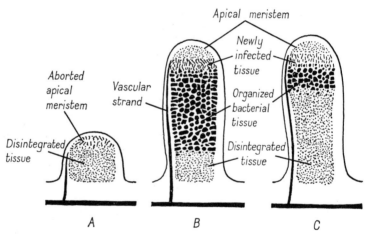

Fig. 5.14. Sections of effective and ineffective root nodules of clover. (A) Ineffective nodule with aborted meristem. (B) Healthy effective nodule. (C) Ineffective nodule with too rapid disintegration. (After Brierley (1957). Plant Symbiosis. *School Science Review*, No. 136, p. 360.)

several factors, such as the genetic constitution of the host legume and of the bacteria, the availability of minerals such as boron in the soil and the actual age of the nodule itself. As the nodule develops, the bacteria multiply, utilizing carbohydrates produced by the host, and the nodule becomes effective and the bacteria nitrogen fixing. At this time the nodule may appear pink due to the formation of haemoglobin and it is now known that the haemoglobin has an important function in the fixation system. Ultimately the bacteria at the base of the nodule attack the host tissues and disintegration results. In some cases the genetic constitution of the host and unfavourable soil conditions may result in the apical meristem not functioning properly or, on the other hand, the rate of disintegration of the tissues may be too fast and very little fixation may take place. The factors governing the formation of healthy and effective nodules are therefore very specific and carefully balanced between host and bacterium.

Nitrogen fixation by symbiotic organisms in association with plants is by no means confined to leguminous plants, and is probably a very much wider phenomenon than is generally realized. The lichens are well-known examples of symbiotic organisms, part fungus and part alga. Recent research has indicated that in some species, such as *Peltigera praetextata*, in which the algal partner is *Nostoc*, there is a significant transference of fixed nitrogen from the blue-green alga to the fungus. It has also shown that many lichens also contain bacteria of the genus *Azotobacter*, but here there is no conclusive evidence that any nitrogen fixed by the bacteria contributes significantly to the overall level of fixed nitrogen in the lichen. The thalloid liverwort *Blasia pusilla* is another organism which contains colonies of *Nostoc*; these are easily seen as small, dark, hollow spots on the lower surface of the thallus. Among the ferns, *Azolla filiculoides*, a small fern found occasionally in southern England floating in ditches and ponds, has also been shown to be symbiotic with the blue-green alga *Anabaena*; as a result the fern is possibly able to live in water deficient of fixed nitrogen.

In higher plants the association of the alder *Alnus glutinosa* with an actinomycete, a small filamentous organism having some affinities with the bacteria, results in the formation of

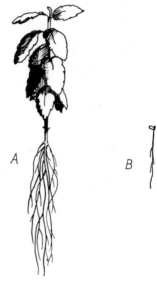

Fig. 5.15. Alder (*Alnus glutinosa*). Growth of nodulated and un-nodulated plants. Grown for 30 weeks in solution without fixed nitrogen.
(A) Plant with nodules.
(B) Plant without nodules.

root nodules in which nitrogen fixation takes place. Experiments using the heavy ^{15}N isotope of nitrogen have indicated convincingly that the nitrogen fixed in the nodules is made available to the whole plant. Just as in the *Rhizobium*–legume relationship, there is evidence that neither the alder nor the actinomycete are able, on their own, to fix nitrogen. The effect of the presence of nodules when the alder is growing on a medium not containing fixed nitrogen is shown in fig. 5.15, on the other hand, if the alder is supplied with fixed nitrogen in the form of nitrate, then the presence or absence of nodules makes little difference to the growth that takes place (see fig. 5.16).

There are numerous other examples of reported and confirmed symbiotic and nitrogen-fixing associations of higher plants with micro-organisms, and it seems likely that such relationships are surprisingly wide-spread, contributing to a very real extent to the amount of fixed nitrogen available to plants and the fertility of the soil.

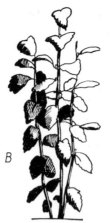

Fig. 5.16. Alder (*Alnus glutinosa*). Growth of nodulated and un-nodulated plants. (A) Nodulated plant grown for 21 weeks in solution containing no nitrate. (B) Un-nodulated plant grown for the same time in solution containing nitrate.

A   B

5.8   The mechanism of nitrogen fixation

The use of ^{15}N has been very useful for research on the mechanism of the fixation process. The most conclusive work, done by Professor P. W. Wilson in America, has been carried out mainly using *Azotobacter* cultures. In one set of experiments it was established that if *Azotobacter* cultures were placed in gaseous ^{15}N, then the isotope was rapidly accumulated by the bacteria. In another set of experiments *Azotobacter* was placed in ordinary gaseous nitrogen, but the bacteria were supplied with ammonia labelled with ^{15}N. Under these conditions *Azotobacter* utilized the ammonia rather than the gaseous source. In both experiments the labelled nitrogen was found in the simple amino-acid 2-aminopentanedioic acid. This suggested that *Azotobacter* takes up atmospheric nitrogen and uses it to form ammonia.

Bergersen and Wilson (1959) investigated the site of the fixation process by centrifuging extracts of soya bean nodules that had been in an atmosphere containing ^{15}N. They obtained three fractions: the host membranes, the *Rhizobium* bacteria and a water-soluble fraction. Initially most of the ^{15}N was in the membrane fraction, but as the experiment progressed more accumulated in the water fraction. Little was found in the bacteria. This suggests that the actual site of the fixation was the host cell membrane. The bacteria must use host photosynthetic products for respiration and so will be effective in producing reducing power. It is thought that the haemoglobin acts as an intermediary in transferring hydrogen from the bacteria to the atmospheric nitrogen so that the latter can be reduced to ammonia. The enzyme nitrogenase, which is responsible for this fixation has been partially characterised, and consists of at least two components, one containing iron and the other both iron and molybdenum. The conditions in the healthy nodule are thought to be relatively anaerobic and this will help to maintain the haemoglobin in its appropriate state for the reduction of nitrogen. The reduction of nitrogen to ammonia requires a source of reducing power and a considerable amount of energy. Reduction is carried out by electron donor molecules such as ferredoxin while the necessary energy comes from ATP, both being derived from respiration (see p. 85). The overall process can be described by the following equation:

$$N_2 + 6e^- + 6H^+ + nATP \rightarrow 2NH_3 + nADP + nP_i$$

in which the $6e^-$ represents the electrons supplied by the donor and P_i the inorganic phosphate produced from the ATP. The value of n is about 12.

The ammonia formed is used in the amination of the acids produced in the respiratory processes of the bacteria and the cell to form amino-acids. For instance the Krebs' Cycle intermediate 1-oxobutanedioic acid combines with ammonia to form 2-aminopentanedioic acid (see p. 105). These reactions are summarized in fig. 5.17.

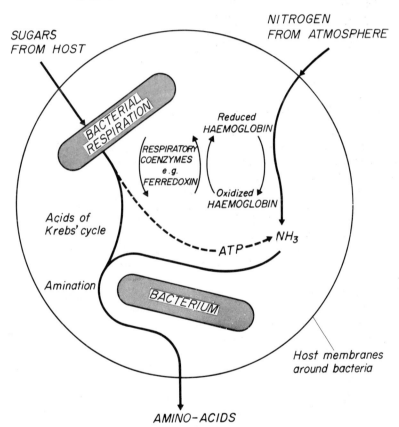

Fig. 5.17. Nitrogen fixation in the root nodule cell. (Modified after Bergersen (1960).)

There have been some interesting suggestions recently relating to the fixation process. If it were possible to produce similar nodules or at least a nitrogen-fixing system in plants like grasses on which man is mostly dependent for food then it would mean that such plants would need little or no nitrogenous fertilizer. Such a situation would depend on some mechanism for modifying the genetic constitution of the cereal by introducing bacterial genes that coded for the missing enzymes. Alternatively the host genes would have to be altered so that bacteria like *Rhizobium* could be accepted and nodules formed. In view of the increasing costs of fertilizers (which take a good deal of energy to produce) such a 'natural' system has enormous economic possibilities in a world where there already is an acute fertilizer shortage.

A final difficulty is how does the fixed nitrogen, now in the form of aminoacids, find its way out into the soil? It is almost certain that both legumes and free-living bacteria secrete reasonable quantities of nitrogenous substances into the soil. The most likely suggestion is that they are secreted in the form of aminoacids which are readily absorbed by plant roots. Other *semi-symbiotic* relationships are currently receiving attention as it is known that some tropical grasses (C-4 plants, see p. 65) 'leak' substantial quantities of carbohydrate into the rhizosphere, the soil surrounding the root, and this can provide a suitable environment for the growth of the free-living nitrogen-fixing organisms such as *Azotobacter*. Research is now going on to find the circumstances that promote this N-fixing association and it has been shown that the genotype of the grass (maize) is important. This suggests that relatively straightforward breeding of a suitable strain could help to maximize this type of activity.

5.9   Some agricultural and horticultural aspects of mineral nutrition

The study of mineral nutrition is obviously of vital importance in agriculture, where crop yields may be considerably increased by mineral addition. This is not always so easy as it may seem. The ideal method for adding minerals is still thought to be by organic manuring, as this provides the plants with a more or less balanced mineral supplement, as well as humus, which will help in the formation of a good soil structure. Unfortunately organic manures are expensive, and often hard to obtain in sufficient quantity, and so inorganic nutrients are nowadays used more frequently. There are several difficulties in the use of these substances. In the first place many of the minor nutrients, being required only in terms of a few parts per million, can easily be provided at too high a concentration and reach a toxic level. Secondly, many (e.g. zinc) tend to become unavailable due to adsorption on to the soil colloids. Finally, if these substances are to be of any use they must be moderately soluble, and this implies that they may be leached or washed out of the soil and lost to the plant. However, if soil drainage water contains too much dissolved fertilizer this may cause serious pollution problems in lakes and rivers. The situation known as *eutrophication* may occur where the mineral-rich water may bloom with algae which, when they decay, cause the water to become deoxygenated.

It is most necessary to add the correct type and quantity of fertilizer for the desired crop. Some plants have higher demands for some nutrients than others. In other cases, although a fertilizer may cause additional growth, this may not occur in the part of the plant that is to be harvested. This can occur in potatoes, which, if given too much nitrogen, make considerable leaf growth and less tubers than they would otherwise have produced. Such a procedure is clearly most uneconomic. Recently some new fertilizer techniques have been tried for the addition of minor nutrients, in particular spraying the leaves rather than the soil with a solution containing the nutrient. Considerable uptake of nutrient may take place both through the stomata and the cuticle, especially where the cuticle is not too thick. For this reason young leaves often take up much more foliar-applied nutrient than older ones. Because of the efficiency with which the nutrient is supplied direct to the plant this technique may be much more economic than applying the fertilizer direct to the soil.

Another technique involves the use of an organic *chelate* of iron from which the metal can be absorbed by calcifuge plants even in the presence of calcium ions. This may be of considerable value to gardeners who wish to grow calcifuges such as *Camellia* and *Rhododendron* on soils of high pH, though this substance is also eventually washed out of the soil and periodic additions are needed.

Finally, it must be emphasized that if crops are to be grown economically to give maximum yield and gardeners are to be successful with their more difficult plants, it is first necessary to understand the particular requirements and preferences of the crop or plant. Secondly, it is necessary to have a knowledge of the mineral status of the soil so that any necessary modification of it may be carried out with the minumum of expense.

Further reading on Mineral Nutrition

BEAR, P. E., *et al.* (1949). *Hunger Signs in Crops.* Washington, American Society of Agronomy.

HEWITT, E. J. and SMITH, T. A. (1974). *Plant mineral nutrition.* London, English Universities Press.

HOAGLAND, D. R. (1944). Lectures on the Inorganic Nutrition of Plants. *Chronica Botanica.*

PEEL, A. J. (1974). *Transport of nutrients in plants.* London, Butterworths.

RUSSELL, SIR E. J. (1957). *The World of the Soil.* London, Collins.

RUSSELL, E. W. (1973). *Soil Conditions and Plant Growth,* 10th edn. London, Longmans.

SCOTT, G. D. (1969). *Plant Symbiosis.* Institute of Biology's Studies in Biology no. 16. London, Edward Arnold.

STEWART, W. D. P. (1966). *Nitrogen Fixation in Plants.* London, Athlone Press.

SUTCLIFFE, J. F. and BAKER, D. A. (1974). *Plants and Mineral Salts.* Institute of Biology's Studies in Biology no. 48. London, Edward Arnold.

WALLACE, T. (1961). *Mineral Deficiencies in Plants.* London, H.M.S.O.

6

The Biochemistry of Cell Activities

6.1 Introduction

In the foregoing chapters some of the basic vital processes that go on both in the cell and the plant as a whole have been analyzed for the most part as separate functions. In this chapter the intention is to try to indicate how the cell acts, at a biochemical level, as an organized entity. To do this we must consider the steps in metabolism which involve the role and reduplication of the nucleic acids, the synthesis of proteins and the working of enzymes, the formation and functions of the more complex carbohydrates and of the fats.

In this way, with the wealth of detail on the ultra-structure of the cell that is now available, we may gain a glimpse of the cell acting as a self-contained, organized whole, but even so, varying in its function within the limits imposed by its own genetic constitution and relative to its position in the scheme of organization of the plant as a whole.

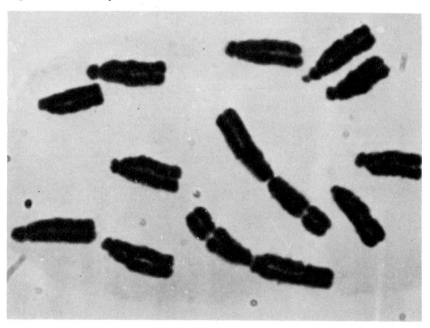

Fig. 6.1. Mitotic metaphase in the broad bean (*Vicia faba*). Note the short, double chromosomes. 2n = 12. (Courtesy Dr C. G. Vosa.)

6.2 The nucleic acids

Since the time of Morgan in the 1910s it has been firmly established that the chromosomes are the site of the hereditary genes, which are arranged in linear order along the chromosomes. In normal cytological preparations of dividing cells (mitotic metaphase) the chromosomes are clearly visible as double, rod-like structures (see fig. 6.1) due to the staining action of the aceto- or fuchsin stains being used (see Appendix, p. 241). These stains are specific for deoxyribose nucleic acid (DNA) found in the nucleus but also in some organelles such as chloroplasts and mitochondria, while the ribonucleic acid (RNA) found in the ribosomes, nucleolus and free in the cytoplasm remains unstained. Other stains such as methyl green pyronin (see appendix 254) can be used to stain both forms of nucleic acid; DNA reacts with the stain to give a green colour, while RNA is stained red.

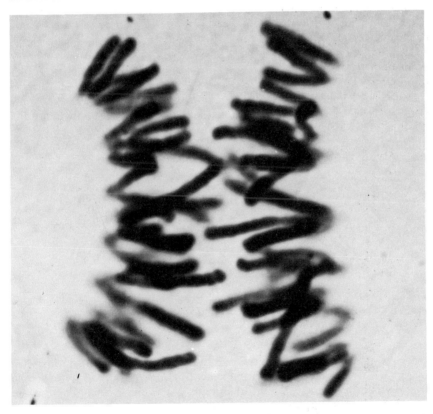

Fig. 6.2. Mitotic anaphase in endosperm of onion (*Allium*). (Courtesy of Dr C. G. Vosa.)

Replication of DNA

There is therefore good evidence that the genes of most organisms are composed of DNA (with the exception of some viruses which contain only RNA).

When mitosis takes place the chromosomes appear as short double strands, the double structure separates at the end of metaphase, and when the chromosomes are last clearly seen at anaphase or telophase (see fig. 6.2) they are single. The problem is, how does the chromosome with its protein core and DNA replicate itself during interphase before the chromosomes appear at the next division?

The answer lies in the unique structure of the DNA molecule. Classical work by Watson and Crick published in 1953 using X-ray-diffraction analysis techniques revealed that DNA is a long complex double helix (see fig. 6.3). Essentially each chain of the helix is composed of deoxyribose sugar molecules joined to one another by phosphate groups. In addition, each sugar molecule has attached a purine or pyrimidine base. This forms a nucleotide chain.

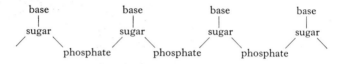

Finally, the two chains are kept together by hydrogen bonding between the opposite bases. These bases are of two kinds, pyrine and pyrimidine, and a purine base on one side of the chain bonds with the opposite pyrimidine base. In this way the four bases of DNA pair so that the purine base adenine pairs with the pyrimidine thymine and likewise guanine pairs with cytosine. Pairing is only possible between these base pairs, *not*, for instance between thymine and guanine. This specific base pairing means that the structure of one helix is effectively determined by the structure of the other, and this fact is crucial in the replication of DNA.

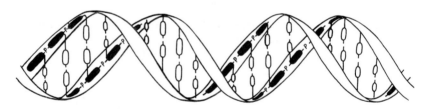

Fig. 6.3. The DNA molecule. Each ribbon represents one nucleotide chain consisting of deoxyribose sugar joined by phosphate groups. The ribbons are held together by H-bonding between opposing bases.

Replication of a complex double helix of this sort, which is coiled upon itself many times to give the visible chromosome structure (see fig. 6.4), must seem a considerable problem. During interphase or, as it is sometimes rather misleadingly called, resting stage, when the chromosome is in a long and relatively uncoiled state, the double helix separates into its two parent chains. This must itself require a progressive separation of the giant molecule of the double helix. As this is effected, then the bonds of the bases are left free for fresh mononucleotides to be attached. The nucleus contains a pool of these mononucleotides (see fig. 6.5), and as the correct, complementary, base is the only one capable of satisfying the particular spare bond, then a new nucleotide chain, identical to the original, is formed and replication is effected. As the

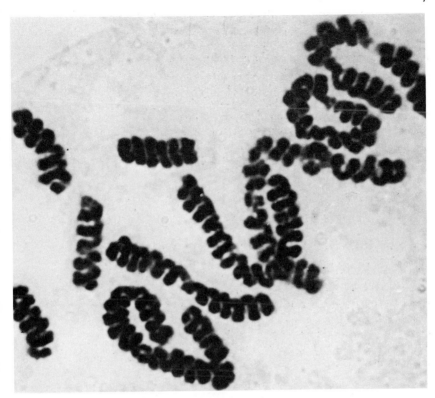

Fig. 6.4. Meiotic metaphase (first division) in the spiderwort (*Tradescantia virginiana*). 2n = 24. Treated with ammonia and alcohol to reveal spiral structure. (Courtesy Dr C. G. Vosa.)

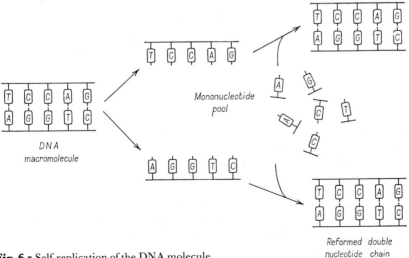

Fig. 6.5 Self-replication of the DNA molecule.

mononucleotides are brought together, so coiling takes place and the original double helix is re-created (see fig. 6.6).

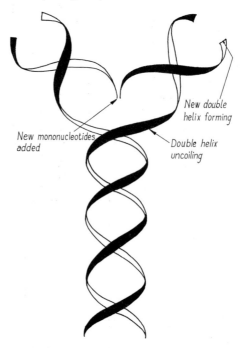

New double helix forming

New mononucleotides added

Double helix uncoiling

Fig. 6.6. Unwinding and replication of the DNA molecule.

An interesting experiment has given support to the concept of a completely new DNA spiral being synthesized on to the template of the other in this manner. The chromosomes of the broad bean are particularly large and clear; if seedlings of the bean are grown in a medium containing the base thymidine labelled with tritium, then on examining the autoradiograph of root-tip metaphase chromosomes, these chromosomes are seen to be uniformly labelled. However, if the radioactive medium is then removed and the next division *but one* examined, then only *one* chromatid of each chromosome is labelled, the other chromatid, newly synthesized in the interphase from non-radioactive thymidine, does not expose the photographic emulsion (see fig. 6.7).

The arrangement of the paired bases along the DNA double helix not only allows for replication of the DNA so that genetic information may be passed on through each cell division but it also provides a blueprint for the synthesis of RNA, a similar nucleic acid, containing ribose instead of deoxyribose sugar and also with the pyrimidine base *uracil* replacing thymine.

The importance of RNA

Analysis of the RNA content of the cell has shown that it is found mainly in the *nucleolus*, the *ribosomes*, the *cytoplasmic* (or *endoplasmic*) *reticulum* and free in the

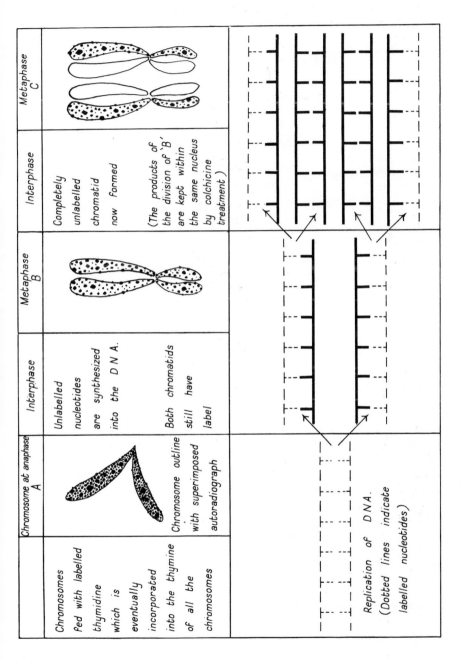

Fig. 6.7. The addition of labelled thymidine to dividing cells.

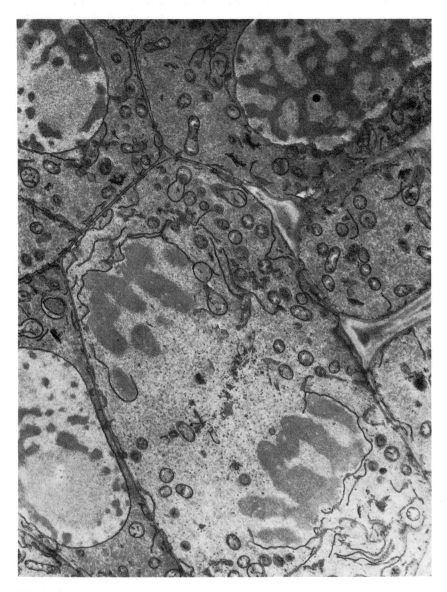

Fig. 6.8A. A telophase cell from the root apex of maize. (See fig. 6.8B opposite.) An electron-microscope photograph showing details of the newly divided cell. Note the two newly formed nuclei with their chromosome material still visible. The nuclear membrane has partially reformed. The mitochondria, cytoplasmic reticulum, ribosomes and *Golgi* body are clearly visible. Plasmodesmata are visible in the surrounding cell wall. (Courtesy of Dr. B. E. Juniper.)

cytoplasm. Electron-microscope photographs (see figs. 6.8 and 6.9) show that the reticulum is a plate-like, folded system, widely dispersed through the cytoplasm. The ribosomes are minute rounded bodies which are usually more or less randomly distributed in young cells, though in older cells they may be grouped into clusters, called *polysomes*. In this state they are often associated with the endoplasmic reticulum which is then often described as *rough endoplasmic*

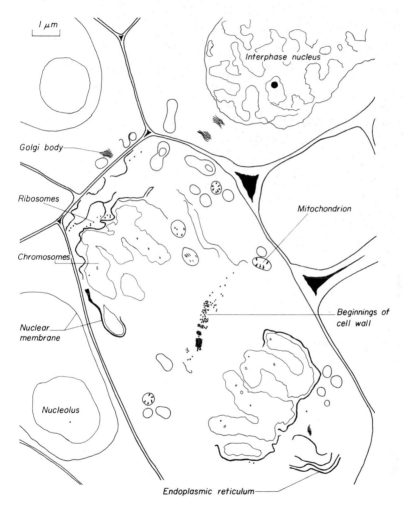

Fig. 6.8B. (See fig. 6.8A opposite.)

reticulum. The *Golgi body* or *apparatus* (see figs. 1.1 and 6.8) appears to be a centre for the organization of the various membranes of the reticulum system and is concerned with the synthesis of a variety of cytoplasmic constituents and cell wall materials. Much of the work carried out on the properties of the ribosomes and reticulum have involved grinding and centrifuging techniques which have

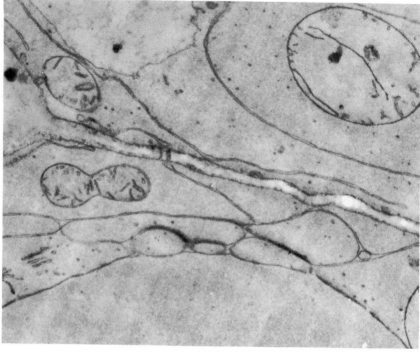

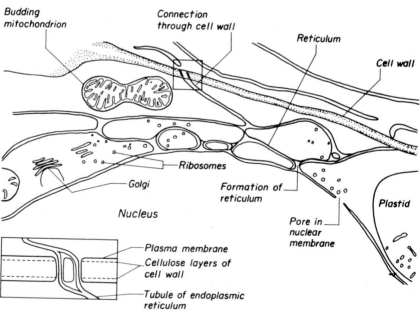

Fig. 6.9. Part of a cortical cell from the root of the broad bean (*Vicia faba*) showing the cytoplasmic reticulum; electron microscope photograph. Insert shows enlargement of part of cell wall with plasmodesmata. (×23,000.) (Courtesy of Dr. B. E. Juniper.)

produced a mass of small particles, partly ribosomes and partly broken-up reticulum. These extracted particles are often given the name of *microsomes*.

Messenger RNA (mRNA)

The DNA template on the chromosome is the blueprint for the synthesis of the *messenger or template* RNA that is found on the ribosomes and possibly to some extent also on the reticulum. Recent electron-microscope photographs have shown that the reticulum frequently appears to connect up with the plasmodesmata or pores in the nuclear membrane (see fig. 6.9). There is a suggestion that this may be a visible sign of some RNA transference from the nucleus to the cytoplasmic organelles. The messenger RNA, organized on to the ribosomes, is the template for protein synthesis (see below and fig. 6.16).

Transfer RNA (tRNA)

The cytoplasmic RNA, called *transfer RNA*, consists of only a relatively small number of nucleotides (about seventy-five), and it is possibly synthesized from other RNA. Its function is also in protein synthesis, as it assists in the conveyance of amino-acids to the correct site on the messenger or template RNA (see fig. 6.17). How the forms of RNA and other substances cooperate in protein synthesis is described below (section 6.5).

Ribosomal RNA (rRNA)

Certain chromosomes have the function of producing RNA that is not primarily concerned in the formation of proteins through the production of a template for their synthesis. This second kind of RNA, referred to as *ribosomal* or *r*RNA, is organized in the *nucleolus* together with proteins to form ribosomes which then pass through the pores in the nuclear membrane and out into the cytoplasm.

6.3 Synthesis of amino-acids and polypeptides

The amino-acids

The chromatograms obtained after algae had been photosynthesizing for some time in $^{14}CO_2$ (see p. 54) showed, in addition to the various carbohydrates, several labelled amino-acids. How these amino-acids came to be formed as indirect photosynthetic products is not entirely clear. On the other hand, use of ^{12}N shows how some amino-acids are produced, for this tracer is rapidly accumulated into 2-aminopentanedioic acid (glutamic acid) through the amination of 1-oxobutanedioic acid (α-ketoglutaric acid), an intermediate of the Krebs' Cycle (see p. 81).

Although about sixty amino-acids have been identified in plant tissues, most cells contain fewer, and only about twenty are concerned in protein formation. Some of these are formed from 2-aminopentanedioic acid by *transamination* reactions. For instance, 2-aminopentanedioic acid and 2-oxobutanedioic acid (oxaloacetic acid) can combine to form 1-oxobutanedioic acid and aminobut-anedioic acid (aspartic acid):

```
        COOH                                    COOH
         |                                       |
        CHNH₂                                   C=O
         |                                       |
2-aminopentanedioic   CH₂                       CH₂         2-oxobutanedioic
    acid               |                         |               acid
                      CH₂              \    /    COOH
                       |                \  /
                      COOH      transaminase
                                   enzyme
        COOH                       system    \
         |                        /            \
        C=O                      /              COOH
         |                      /                |
1-oxobutanedioic  CH₂    ↙         ↘           CHNH₂
    acid           |                            |
                  CH₂                           CH₂       aminobutanedioic
                   |                            |              acid
                  COOH                         COOH
```

Peptide bond formation

When several amino-acids are brought together on to the template RNA they join together by means of *peptide bonds*, to form proteins. The system by which two amino-acids join together to form a di-peptide is illustrated by the synthesis of glutamylcysteine.

```
  COOH              ATP      ADP      COOH
   |                 _____↗         |
  CHNH₂                                CHNH₂
   |                                    |
  CH₂                                   CH₂
   |              ──────────→           |
  CH₂       COOH                        CH₂
   |         |                           |              COOH
  COOH  +   CHNH₂                 ┌─────────────┐        |
            |                     ¦ CO — NH ──¦─CH
           CH₂SH                  └─────────────┘        |
                                                        CH₂SH

2-aminopentanedioic (glutamic) acid + cysteine      glutamylcysteine
                                                        + H₂O
```

This condensation, in which the carboxyl group of the 2-aminopentanedioic acid reacts with the amino-group of the cysteine, requires energy in the form of ATP. The tri-peptide glutathione is formed by a similar condensation reaction with glycine as the additional amino-acid. This energy-requiring, multiple condensation process by which amino-acids are joined together by peptide bonds to form polypeptides gives a repetitive structure of the following form, though the chain is a three-dimensional one with the bonds linking the amino-acids to the chain arranged at approximately 120° to each other. R_1, R_2, R_3 represent these different amino-acids:

```
            R₁                              R₃
            |                               |
           CH        NH         CO         CH
          /  \      /  \       /  \       /  \
       H₂N    CO   CH    NH   CH    COOH
                    |
                   R₂
```

Polypeptides may contain a number of different amino-acids, but each may be represented more than once. The properties of the particular polypeptides—and proteins as well—depend on the identity and the sequence of acids along the chain.

The analogy with the carbohydrates, in which multiple condensation takes place to form long-chain molecules, is not entirely valid, as in the polypeptide chain the units of the chains are usually different, but nevertheless arranged in a specific series. How they come to take up this pattern and themselves become bonded together into even longer chains is a more difficult problem to solve.

X-ray analysis of structural proteins has shown that the amino-acid residues are arranged on alternate sides of the chain. These residues may be either acidic or basic (e.g. aminobutanedioic and 2-aminopentanedioic acids are acidic, lysine and arginine basic), this leads to the possibility of linkages between two nearby chains. Links between amino groups and carboxyl groups are also found and are of a simple electrovalent type involving the removal of a positively charged hydrogen ion (proton) from the carboxyl group to the amino group:

Other commonly found linkages occur through sulphur atoms and through hydrogen bonding between hydrogen atoms and oxygen atoms:

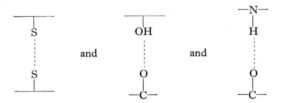

6.4 The organization of proteins and enzymes

There are three main groups of proteins, the *structural proteins*, the *reserve proteins* and the *enzymes*. The first group includes those that make up parts of the various cell organelles, the 'backbone' of the chromosomes on which the nucleic acids are arranged, the structural material of the mitochondria and some of the other cell organelles.

The reserve proteins are not used much as respiratory sources; they are easily broken down into peptides and amino-acids, which are then translocated and metabolized into new structural or enzymatic proteins. For details of the analysis of proteins and amino-acids in germinating seeds and other tissues, see Appendix, p. 241.

The enzymes

The study of enzymes, the organic catalysts of living things, must rate as one of the most important aspects of biology. One of the enigmas of the subject is how these various catalysts can be formed and exist so close side by side within the cell and yet work only at the appropriate place and time required by the general processes of metabolism. Perhaps part of the difficulty is the problem we have in appreciating the scale of the molecular dimensions we are considering, yet study of electron micrographs has shown the discreteness of the organelles and indeed many enzymes can be regarded as being *site-specific* and are an integral part of the mitochondria, chloroplasts, ribsomes and other organelles. Others such as the lysozymes may be enclosed in special membranes where they may be effectively kept away from other cell constituents unless the cell is damaged or becomes disorganized. Many others probably exist, at least from time to time, free as minute colloidal particles in the cytoplasm.

Catalysts are usually defined as substances that change the rate of a chemical reaction without themselves becoming chemically changed in the process. This is true of such inorganic catalysts as platinum or carbon whose activity depends on some relatively simple surface action. The more *labile* or easily changed nature of proteins renders enzymes more susceptible to alterations during the course of a catalysed reaction, so that some regeneration may be required. They are also more specific in the physical conditions needed to carry out effective catalysis. In addition their complex shape and structure makes enzymes much more specific in carrying out a particular reaction than inorganic catalysts.

As in inorganic catalysis most enzymatic reactions are reversible and it is often difficult to understand how the same enzyme may affect a reaction in two directions. A good example of this is in the action of the enzyme triose-phosphate dehydrogenase. In photosynthesis this enzyme co-operates with NAD to reduce PGA to triose (see p. 62), while in respiration triose may be converted into PGA in the opposite manner. The rate at which catalysis may take place is exceedingly variable; some enzymes, e.g. catalase (which breaks down hydrogen peroxide into oxygen and water), work exceedingly fast, and each molecule of enzyme may be able to act on more than a million molecules of substrate in a minute.

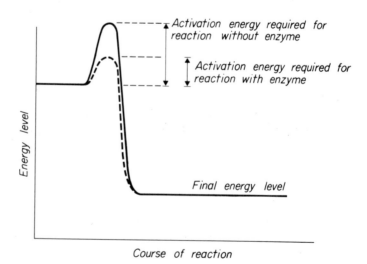

Course of reaction

Fig. 6.10. Activation energy.

Other enzymes may have lower *turnover numbers*, being able to catalyse the reaction of only a hundred or so molecules in a minute.

Activation energy

For most reactions to proceed the substrates have to have their energy increased; this is called their *activation energy*. In the laboratory the activation energy is often increased, and reactions thus made to take place more rapidly, simply by heating the reaction mixture. This clearly does not happen in the living plant yet somehow enzymes are able to lower the activation energy required. This is illustrated in fig. 6.10.

Enzyme-substrate interaction

Some insight as to how enzymes may work has come from studies designed to show how the enzyme may combine with its substrate, albeit temporarily. At the end of the nineteenth century Emil Fischer suggested that an enzyme could have a particular region, an *active centre* which could act in a way analogous to a lock and key. An example of this has already been discussed in the conversion of butanedioic acid (succinic acid) to *trans*-butanedioic acid (fumaric acid) in the

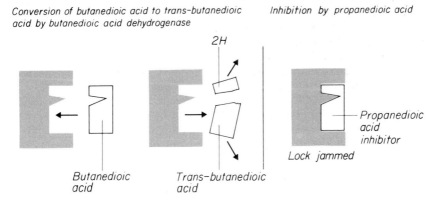

Conversion of butanedioic acid to trans–butanedioic acid by butanedioic acid dehydrogenase

Inhibition by propanedioic acid

2H

Butanedioic acid

Trans–butanedioic acid

Lock jammed

Propanedioic acid inhibitor

Fig. 6.11. Lock and key analogy for enzyme action.

Krebs' cycle (see p. 80). Here the inhibitor propanedioic acid (malonic acid) which resembles the normal substrate very closely, acts as a competitive inhibitor and may be thought of as 'jamming the lock' of the butanedioic acid dehydrogenase enzyme (see Fig. 6.11).

Enzyme-substrate combination

Evidence that the enzyme may actually combine with its substrate has come from a number of sources. For instance, in investigating the formation of phosphate esters of sugars, where glucose-1-phosphate is converted to glucose-6-phosphate, it has been possible using glucose-1-phosphate labelled with ^{32}P to extract the enzyme concerned together with the ^{32}P to which it has become temporarily attached. The following diagram summarizes the kind of system that probably operates in many such enzyme-catalysed reactions:

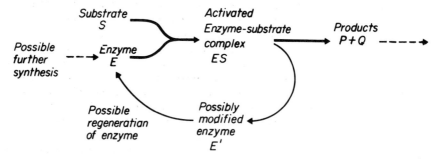

Enzyme specificity

The lock and key model implies that enzymes must be highly specific and only capable of 'fitting' with a few substrates of the right shape. In practice there are some enzymes which are highly specific. Some show *absolute specificity*, for instance catalase, which is only able to catalyse the breakdown of hydrogen peroxide. Others such as the lipases are much less specific and are capable of catalysing a series of reactions with different substances because they all contain a particular bond or group. Because of their three-dimensional shape enzymes are mostly highly stereo-specific, for instance β-glycosidase hydrolyses β- but not α-glycosides to form alcohol and sugar, though the enzyme will attack a wide range of β-glycosides containing differing alcohol groups. Recently much attention has been paid to the phenomenon of *allostery* ('other shape') where the active site or some other part of the enzyme is blocked or modified in some way, not by a substrate-like competitive inhibitor but often by some further product of metabolism. This may be very important in causing feedback inhibition of the enzyme. This would provide a system for regulating the rates of metabolic processes. Conversely the removal of an allosteric binding agent would result in an increase in enzyme activity.

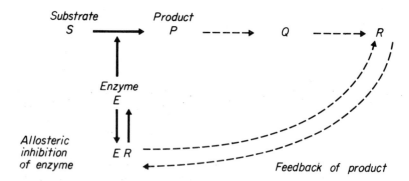

The structure and properties of enzymes

Enzymes can be extracted by grinding the tissue concerned in cold buffer and centrifuging, but the extracts produced will be mixed and impure. Further purification can be achieved by a variety of techniques including further centrifuging and filtration. More or less pure crystalline enzymes can be precipitated from solution using organic solvents such as ethanol and propanone (acetone). The structure and properties of the enzymes can then be examined. Some enzymes are proteins on their own but others consist of a main protein part together with a non-protein or *prosthetic* group which can be separated from the protein by dialysis. Coenzymes are prosthetic groups that are readily dissociated; two common coenzymes, nicotinamide adenine dinucleotide (NAD) and flavin adenine dinucleotide (FAD) are associated with many different enzymes (see p. 83). Enzymes may also include metallic *activator groups* or *cofactors* such as iron and copper in the respiratory cytochromes and ascorbic acid. Some inhibitors, e.g. cyanide, act through forming stable complexes with these metallic groups. Knowing that proteins contain such a great range of amino-acids and that they can exist in such a variety of shapes makes the problem of sorting out the structure of enzymes an exceedingly difficult one and we often have to be content with a gleaning information about their structure from observation of their properties.

Temperature and enzyme activity

Like most chemical processes, enzyme-controlled reactions are temperature sensitive; at relatively low temperatures the reaction will show a Q_{10} of about two which is typical of most chemical reactions, but unlike many inorganic catalysts enzymes have a fairly low *optimum* temperature above which the rate of

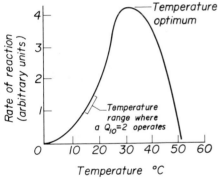

Fig. 6.12. Effect of temperature on the activity of the enzyme polyphenol oxidase.

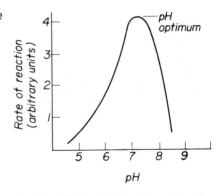

Fig. 6.13. Effect of pH on the activity of the enzyme urease. (After Harrison D., *Patterns in Biology*. (1975) London, Edward Arnold.)

the reaction falls off as thermal inactivation is noticed and the protein is eventually denatured at temperatures approaching 60°C. (fig. 6.12). Different plants have different temperature optima for various metabolic processes such as germination, and it would be interesting to discover the mechanism of adaptation to life at very high temperatures, such as may be encountered by some micro-organisms in hot springs. It is worth noting that some ordinary plants may also experience very high temperatures for short periods when exposed to direct sunlight in open situations and must therefore be adapted to tolerate these conditions.

pH and enzyme activity

Enzymes are usually only capable of working over a fairly narrow pH range and it is perhaps not surprising to find that most cells maintain their pH within fairly narrow limits usually just below neutrality, though there may be considerable diurnal fluctuation in some plants such as the desert succulents. The reason for enzymes being pH sensitive lies in the fact that the amino-acids of which they are composed behave as *zwitterions*; that is they are polar molecules which can accept a hydrogen ion at low pH or donate one at high pH. Thus changes in pH, which will raise the concentration of hydrogen ions at low pH and depress them at high pH, will result in distortion of the enzyme molecule as tensions are set up between individual amino-acids. The effect of pH on the enzyme urease is illustrated in fig. 6.13.

Relationships between enzyme and substrate concentration

As would be expected, provided that the concentration of the substrate is not limiting, the more enzyme that is available, the more substrate that is converted. Similarly, for a given amount of enzyme the amount of substrate converted and the rate of conversion will depend upon the concentration of substrate present, but only up to a certain limit. At a particular concentration of substrate it is likely that all the active sites on the enzyme are fully occupied, and the rate of conversion then reaches a maximum. These relationships are illustrated in figs 6.14 and 6.15.

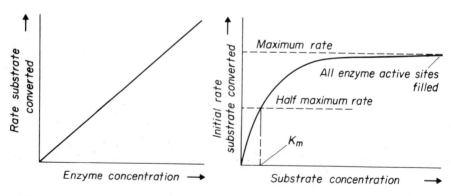

Fig. 6.14. Relationship between enzyme concentration and rate of reaction.

Fig. 6.15. Relationship between substrate concentration and rate of reaction.

Examination of the second graph for a particular enzyme has proved useful in determining the degree to which the enzyme forms a complex with its substrate. The substrate concentration at which half the maximum rate of reaction is achieved is an important value called the *Michaelis constant* (K_m on the graph).

Low values for Km (even as low as 10^{-4}) suggest that the enzyme combines very firmly with its substrate while high values approaching one mean that there is only a very loose affinity between the two (for further details of this aspect of biochemistry you are advised to consult the biochemical texts listed at the end of this chapter).

No discussion of enzymes and their importance should be complete without emphasizing the great variety and importance of the group in almost all anabolic and catabolic reactions in plants, animals and micro-organisms (see table below). It is also interesting to note that very similar enzymes occur in all groups of living things; perhaps another example of a unifying theme in biology.

Group	*Function*	*Examples*
1. Oxidoreductases	Oxidation and reduction reactions, e.g. in respiratory processes.	Butanedioic (succinic) acid dehydrogenase. Triosephosphate dehydrogenase. Polyphenol oxidase. Ascorbic oxidase.
2. Transferases	Various transferring reactions.	Transaminase Phosphokinase.
3. Hydrolases	Often in catabolism in which water molecule is introduced to break links such as erte, peptide or glycosidic bonds.	Peptidases Amylases Sucrase Maltase Lipases Urease
4. Lyases	Removal or addition of groups to create or break double bonds.	Decarboxylases Deaminases.
5. Isomerases	Changing shapes of molecules.	Phosphoketoisomerase.
6. Ligases (synthetases)	Linkage of two molecules, coupled with breakdown of ATP or other triphosphates.	Aminoacyl-tRNA synthetases.

Table of the main groups of enzymes.

6.5 Protein synthesis

In protein synthesis, RNA provides the blueprint or template on which the synthesis takes place. Amongst the evidence for this is that from experiments in which cells or extracts from cells are treated with the enzyme ribonuclease. This enzyme, which breaks down ribonucleic acid, also usually inhibits the formation of new protein. Use of radioactive ^{14}C labelled amino-acids indicates that peptides are themselves not particularly important in protein synthesis, as they seldom accumulate in more than trace amounts. On the other hand, it is possible that they are immediately converted into protein as they have a high turnover. However, the most likely conclusion must be that the amino-acids are joined together on one surface to form the protein direct. This occurs by means of the rapid sequential formation of peptide bonds.

Evidence regarding the site of this synthesis has come from the use of ^{14}C together with fractional centrifuging techniques. Amino-acids containing ^{14}C tracer are injected into a living tissue, which is then killed, ground up in isotonic ice-cold buffer and centrifuged slowly, at medium speed and then at very high speeds of about 100000 $\times$g. It is found that only the RNA-rich microsomal fraction, thrown down at the last high speeds, contains the tracer, and the tracer is mostly incorporated into proteins. If the sampling is carried out soon after

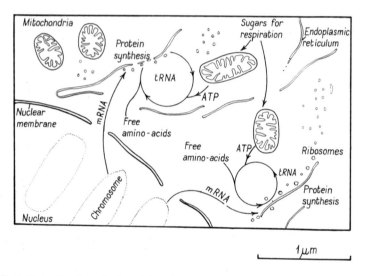

Fig. 6.16. Co-ordination of cell activities in protein synthesis.

application of the radioactive amino-acids, then the ribosomes carry the tracer, but if the sampling is delayed for some time, then they are no longer radioactive. In other words, a fresh series of unlabelled amino-acids has been metabolized, the first having been released into the cytoplasm as proteins.

In the neighbourhood of the endoplasmic reticulum and ribosomes (see fig. 6.16) there are already present free amino-acids, tRNA, ATP and certain enzymes. It is thought that an enzyme forms a surface for bringing these together, ATP providing the energy so as to form a complex of the enzyme, tRNA and the *appropriate* amino-acid (see fig. 6.17). The other form of RNA, mRNA, which is assembled on the DNA template in the nucleus, also diffuses out of the nucleus, and takes up a position on the ribosome, where it forms a template for the protein synthesis. The smaller tRNA complexes then arrange themselves alongside the larger mRNA template, obtaining the right 'fit' in the process. The recognition or coding part of the mRNA is termed a *codon*, that part of the tRNA that is able to fit with it is called the *anti codon*. In this way the correct amino-acids required for a particular protein or enzyme are brought together *in the correct sequence* as determined by the mRNA template, the structure of which is itself determined by the original nuclear DNA of the gene.

Finally the bond joining the amino acid and the RNA is broken and new peptide bonds are formed between each amino acid. As this is accomplished the chain of amino-acids, now a polypeptide, leaves from the transfer RNA and the latter is released for further synthesis or is destroyed (see fig. 6.17).

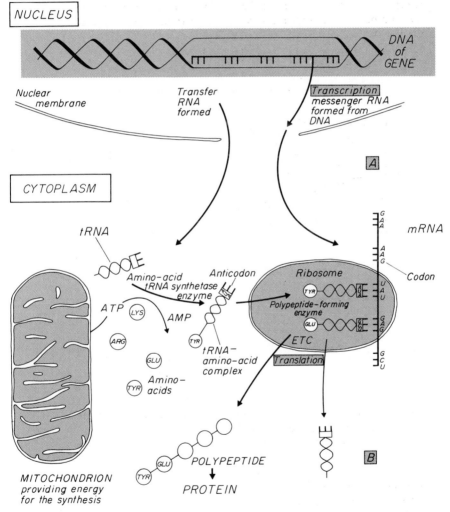

Fig. 6.17. Diagram to illustrate the mechanism of protein synthesis. (A) *Transcription*: The nucleotides of the DNA are transcribed to mRNA. (B) *Translation*: An amino-acid, tRNA complex finds a site for its *anti-codon* with the appropriate *codon* of the mRNA. Correct sequences of amino-acids are thus put together to form a polypeptide and protein. (Modified after Strickberger, M. W. (1968) *Genetics*, London, Macmillan.)

This mechanism for protein synthesis—involving a transference of information from the nucleus to the ribosome site in the form of mRNA and requiring the co-operation of the mitochondria—emphasizes that the cell, to work efficiently, or indeed at all, must respond as an organized and integrated whole. The recent advances in biochemistry and electron microscopy have made it possible, for the first time, for us to visualize how at least some of these vital processes take place.

The genetic code

The year 1961 can be regarded as a milestone in the study of cell organization, as the work of Crick at Cambridge and Nirenberg in the United States has shown us something of the mechanism by which *different proteins* are produced by a given nucleotide complement on the chromosome. In other words, they have indicated some of the first steps in the sequence of *how genes act.*

Nirenberg's work has been based on the technique of extracting the ribosomes, feeding them with mRNA of a known base sequence and then analysing the amino-acid order of the protein produced. Analysis of RNA has shown that it can contain four bases, adenine, guanine, cytosine and uracil. If mRNA consisting of only *uracil* nucleotides is fed to the extracted ribosomes in the presence of the twenty or so amino-acids that are commonly found in the cytoplasm, then a protein is formed consisting of only one amino-acid type, *phenylalanine.* Similarly, a strand of mRNA composed of only *cytosine* groups produces a protein with only *proline* amino-acid units.

The next problem is *how many* nucleotide bases of the messenger RNA are required to order one amino-acid? With only four bases to select in the code and twenty amino-acids to form, two nucleotides would be too few, as only sixteen combinations can be produced by these bases. Three nucleotides might seem a likely number. Another problem relates to the reading of the code for the whole nucleotide sequence. Is there any indication where a set of the three nucleotides

Scheme to show that the reading of the genetic code of nucleotides depends on reading the whole sequence from one end (Hypothetical DNA nucleotide sequence)

A=adenine, G=guanine, C=cytosine, T=thymine

begins, or does the code simply start at one end of the nucleotide series and simply 'read along' counting off three nucleotides at a time?

The answers to these problems have come from Crick and his co-workers' studies on a virus which attacks bacteria, called the T4 bacteriophage. This was treated with proflavine, a chemical mutagen that causes deletion or addition of a nucleotide base, so that it would not attack one of two strains of bacteria. By careful use of the mutagen they were able to show that the code reads from one end of a DNA chain. First a mutation was caused near the start of the nucleotide sequence; this mutant will only attack one strain of bacteria. A second mutation was caused in the bacteriophage close to the site of the first mutation which would 'displace the type' of the code so that the effect of the first mutation was removed (see above). Mutations farther along the nucleotide sequence would still not be fully viable, as the order of nucleotides has been upset by the first mutation, near the start of the sequence. The new mutation must be near the first if the correct order of sufficient nucleotides is to be restored.

Similarly, the virus could be made to change back to the normal type by introducing *two more mutations*, both of which add another nucleotide to the sequence.

II A A G* | G C T | T G A | C C A | T →*mutant* will not attack one
                                                strain of bacteria

    start⟶

            Treated with mutagen
IV A (C) (C) | A G* G | C T T | G A C | C A T →*near-normal*, which
                                                will again attack
                                                both strains of
                                                bacteria

    (2 new nucleo-   triplet sequence
    tides added)   now restored

            Original normal
I   A A G | C T T | G A C | C A T →

Scheme to show that the reading of the genetic code is carried out in groups of three nucleotides at a time
The sequence of nucleotides in each triplet is the same in the final mutant (IV) as in the normal (I)

The code of nucleotides on the messenger or template RNA that determines the selection of a particular amino-acid for incorporation into a particular protein is thus at least partially understood, and we are much nearer an understanding of how a particular gene produces a particular protein. However, many problems remain; we have little idea how the system is integrated and how different instructions may be given as cell-division and differentiation proceed. Nevertheless, the decoding of some steps in the sequence of protein synthesis must remain an important step in the history of biology.

6.6 Carbohydrate chemistry

In Chapter 3 we discussed the formation of the main product of photosynthesis, fructose-diphosphate. This substance is converted into a wide range of structural and energy-reserve carbohydrates and carbohydrate derivatives which are of great importance in the organization of the cell.

Structure of sugar molecules

There are two outstanding features of carbohydrate conversions that must be considered from the outset, first, the isomeric properties of the hexose unit, and secondly, the conversions which take place involving the phosphate ester of the particular sugar.

Fructose, for example, can exist in two isomeric forms, one being the six-sided pyranose ring, which is more usual, the other being the more reactive, less stable, five-cornered furanose form:

fructopyranose   fructofuranose

The numbers indicate the standard numbering of each carbon atom. In addition to these two isomers there are also *stereoisomers*. In these there is only a spatial difference; for example, glucopyranose can exist in two forms, α- and β-glucopyranose. These can be distinguished by their differing abilities to rotate the plane of polarized light (see notes on the use of the polarimeter in Appendix, p. 243). Both are dextro-rotatory, but the α- form more strongly than the β-form:

α-glucopyranose   β-glucopyranose

These various isomeric forms are of considerable importance, as in the combination of fructofuranose and glucopyranose in sucrose synthesis; in the formation of polysaccharides, by condensation of α-glucopyranose units giving starch. Similarly the condensation of β-glucopyranose units gives cellulose.

Attention has already been drawn to the formation of phosphate esters of sugars in both photosynthesis and respiration; few carbohydrate conversions take place unless the hexoses are in the form of the phosphate esters. This is probably because many reactions require the co-operation of enzymes capable only of working on the phosphate esters. Looked at another way, one could say that the formation of the phosphate ester from the ordinary hexose sugar requires high-energy phosphate (ATP), and this, through the addition of the phosphate radical, is necessary for the eventual condensation. The final product of photosynthesis may be regarded as fructofuranose-diphosphate. A vast range of more complex substances are made from this substance as the essential starting-point; the reserve carbohydrates, the structural carbohydrates, the vacuolar pigments and the glycosides. As most of these are vitally important in the cell, it is necessary to know how and where they are formed.

6.7 The reserve carbohydrates

The two most important of these carbohydrates, which are easily utilized as food and energy sources, are the disaccharide sucrose $(C_{12}H_{22}O_{11})$ and the polysaccharide starch. In many plants these are formed quickly and directly from the fructose produced in photosynthesis.

Sucrose

Sucrose is formed by a combination of fructofuranose-6-phosphate with glucopyranose-6-phosphate:

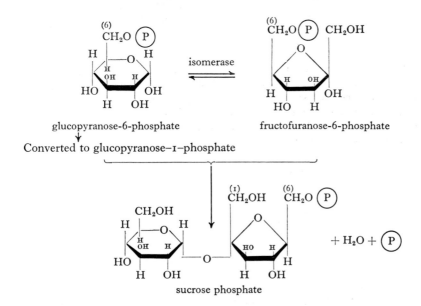

glucopyranose-6-phosphate

fructofuranose-6-phosphate

Converted to glucopyranose–1–phosphate

sucrose phosphate

This synthesis requires a condensing enzyme and uridine-triphosphate. The overall reaction requires energy derived from the conversion of ATP to ADP.

The sucrose can be hydrolysed to form glucose and fructose by the enzyme invertase or quite simply in the laboratory by the action of dilute acid. As these products have different optical properties (glucose being weakly dextro-rotatory, fructose laevo-rotatory), the activity of the enzyme invertase can be examined in the polarimeter (see Appendix, p. 243). The sucrose is strongly dextro-rotatory, and the final product or invert sugar mixture is weakly laevo-rotatory. Although some plants, particularly sugar beet and sugar-cane, store large quantities of sucrose, it is not usually the primary reserve carbohydrate. Owing to its solubility and consequent ease of transport, it is found in the cytoplasm of many cells and may be transported considerable distances (for instance, to the nectaries of flowers) through the phloem cells.

Starch

In all essentials starch is probably a long-chain condensation product which can be formed from glucopyranose-1-phosphate molecules by the action of starch phosphorylase (though other enzymes are also capable of effecting the synthesis).

Synthesis of amylose by the formation of 1–4 α links between glucose molecules

Synthesis of amylopectin by the formation of 1–6 α links between glucose molecules

This simple starch involving 1–4 α links between the glucose units is called amylose. This reaction can be carried out in the laboratory using an enzyme extract obtained from potato (see Appendix, p. 245). Starch also consists of a second substance, amylopectin, in which there are 1–6 α links, as well as the 1–4 α links; the former are produced due to the activity of a so-called *branching enzyme*.

In amylopectin each branch may consist of between twenty and thirty glucose units, but the overall macromolecule may have been formed from hundreds of hexose molecules.

Starch must generally rate as the most important storage carbohydrate in plant cells, and small grains of it are usually visible in the chloroplast itself soon after photosynthesis has begun in an illuminated leaf. Accumulated starch in storage organs, such as rhizomes and tubers, is another problem, as it must necessitate the conversion of the starch into soluble material that can be easily translocated both to and from the storage organ. The starch is converted into soluble sugars by the action of a group of hydrolysing enzymes called amylases.

6.8 The structural carbohydrates

The carbohydrates also play a vital role in plants in providing their cells with strength. The simplest of these structural substances is cellulose, but in some tissues this may become replaced by the more complex lignin.

The cellulose macromolecule is superficially similar to that of amylose, but β-glucopyranose units are joined by 1–4 β links. This type of linkage results in extremely long-chain macromolecules which may contain as many as 2500 glucose residues. Additional strength is given to the molecule by hydrogen bonding beyween the hydroxyl groups of the different chains; in this way the cellulose takes up a crystalline form and chains of great tensile strength are built up. The arrangement of these chains in the cell is of considerable interest. The middle lamella, that is the layer between the primary walls of two adjoining

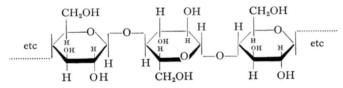

layers, is composed of a mixture of calcium and magnesium pectates (see below) and forms a non-crystalline, colloidal layer. The primary wall is the next to be

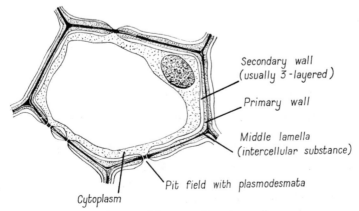

Fig. 6.18. The cell wall constituents of the parenchymatous cell.

formed by the growing cell, and consists mostly of cellulose, together with various pectic compounds. As the cell grows various cellulose layers are laid down, and

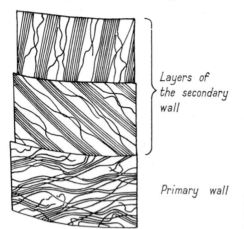

Layers of
the secondary
wall

Primary   wall

Fig. 6.19. Arrangement of microfibrils of cellulose in the outer layers of the cell wall.

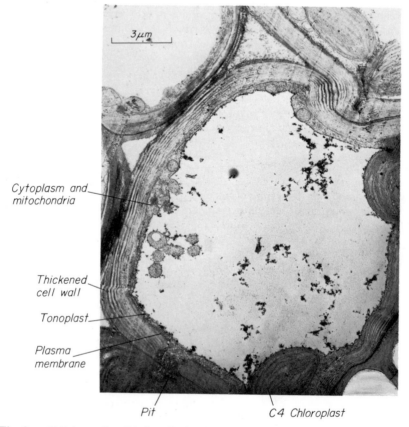

Cytoplasm and
mitochondria

Thickened
cell wall

Tonoplast

Plasma
membrane

Pit   C4 Chloroplast

Fig. 6.20. Cellulose cell wall in bundle sheath of sugar cane (*Saccharum*) leaf.

this may result in the formation of obvious laminations. Finally, when the cell has ceased to enlarge, the formation of the secondary cell wall may take place, and this, like the primary wall, also consists mainly of cellulose (see figs. 6.18, 6.19 and 6.20).

Cellulose serves a vital role in the cell in enclosing the whole protoplasm and acting as a strong, yet fully permeable, elastic boundary wall.

Pectic substances

An important group of carbohydrate derivates that are often found impregnating the cell wall and even in the cytoplasm of the cell are the pectins or pectic substances. These are also polymers, but are derivatives of the sugar, galactose, through the formation of galacturonic acid. Residues of galacturonic acid are joined together by series of condensations like those found in the formation of starch and cellulose.

Galacturonic acid units

pectic acid

The pectic acid chain will then react with calcium and magnesium ions to form insoluble pectates. These divalent ions act as bridges between separate pectic acid chains and so add to the strength of the whole.

6.9 Lipid (fat) metabolism

Although the carbohydrates tend to be regarded as the most general and important of the food-reserve substances in plants, this is sometimes an erroneous view, engendered perhaps by our use of cereal crops for food. In most plants there are substantial quantities of lipids or fats included in the seed reserves. In addition, these may accumulate in young tissues and in the leaves, for instance large fat globules are visible in the stomatal guard cells of the privet, and many fungi and algae use lipids as their main food reserve. Phospholipids are an important group of fats that combine both water-loving (hydrophilic) and water-repellant properties. They are often found in the cell-membranes, the plasma membrane and the tonoplast, where their bipolar structure gives the membrane its characteristic properties.

Structurally, fats and oils are derivatives of glycerol and fatty acids. Glycerol is a three-carbon alcohol which is derived from three carbon (triose) sugars:

A wide range of fatty acids is formed in plant tissues, but the commonest include palmitic acid, stearic acid and oleic acid, e.g.

$$CH_3(CH_2)_{14}COOH$$ Palmitic acid

The formation of a fat is essentially one of esterification through the union of the acid with the alcohol:

$$
\begin{array}{l}
CH_2OH \\
| \\
CHOH + 3 . C_{15}H_{31}COOH \longrightarrow \\
| \\
CH_2OH \\
\text{glycerol} + \text{palmitic acid} \longrightarrow
\end{array}
\qquad
\begin{array}{l}
CH_2O(C_{15}H_{31}CO) \\
| \\
CHO(C_{15}H_{31}CO) + 3H_2O \\
| \\
CH_2O(C_{15}H_{31}CO) \\
\text{palmitin} + \text{water}
\end{array}
$$

The second group of liquids, the waxes, differ from fats and oils in that the glycerol is replaced by monohydric or occasionally dihydric alcohols. These substances are important in that they line the epidermis of a wide range of plants, lowering cuticular transpiration. Electron-microscope photographs of the surfaces of leaves show some of the beautiful arrangements that these waxy layers may form (see fig. 2.12 on p. 23). Cutin and suberin are wax-like substances also found in epidermal layers and are the condensation products of various fatty acids. Such substances also occur in the Casparian strip of the endodermis, where they are important in preventing the movement of water and solutes through the cell walls. Details of simple tests for lipids are given in the Appendix, p.247.

The pathway for the formation of the fatty acids is complex, but it is thought that they are formed from acid intermediates such as acetyl-coenzyme A, the acetic acid derivative formed at the start of the Krebs' Cycle (see p. 79).

6.10   Phenolic Substances

These are common and important substances often found in immense quantity in plants. They occur in great variety but can be broadly divided into two main groups; those based on a single benzene ring and those formed essentially from three units, two of which are benzene rings. The first group includes the phenolic acids proper, acids that are the main constituents of the *lignin* of woody tissues; the second group includes the *anthocyanidin pigments* that form the blue and red pigments of flowers and leaves. This group, as a whole, is called the *flavonoids*.

Coumaric acid,
a constituent of lignin

Pelargonidin,
a red flower pigment

Lignin

Under certain circumstances, for instance, in fibres, sclerenchymatous cells and developing xylem vessels and tracheids, parts of the cell wall may become progressively impregnated with a complex aromatic substance; the phenolic derivative *lignin*. It may be present in all three cell wall layers. Staining with phloroglucinol and HCl, which reacts with lignin to give a bright magenta-red colour (see appendix p. 245), shows that lignification usually starts from the outside of the cell, and in heavily lignified areas it is usual to find the middle lamella and primary wall most strongly thickened. Lignin is therefore an important constituent of woody tissues, making up between twenty and thirty-five per cent of the solid material and thus contributing enormously to the strength and rigidity of the plant structure.

The mechanism of formation of lignin is complex, but is thought to involve the formation of *shikimic acid* from carbohydrates. After a series of further conversions the aromatic amino-acid *phenylalanine* is formed. The next step is an important one and is known to be catalysed by the enzyme phenylalanine ammonia lyase (PAL) to form *cinnamic acid*. PAL is known to be sensitive to certain environmental factors such as light (possibly mediated through the phytochrome system, see p. 189). Further conversions allow for the formation of the main lignin precursors coumaric, ferulic and sinapic acids. These become linked together as a complex polymer that is both rigid and resistant to decay.

Shikimic acid   Phenylalanine   Cinnamic acid

It is well-known that many phenolic substances are toxic and it may be worthwhile speculating on their origin. It is possible that they represent waste products of metabolism that would be dangerous to the plant if allowed to accumulate, accordingly the deposition of such substances would prove to be an adaptation of considerable survival value not only in terms of providing strength, but also in removing potentially toxic substances from areas of active metabolism. Indirect advantage is also gained as these substances are difficult to utilize as food sources and are only digested by a relatively small group of organisms, principally the fungi and bacteria. Old woody tissues, particularly of the heart wood which is further impregnated with such phenolic substances as the tannins is likely to be even more resistant to decay, though this is probably due to the bitterness and unpalatibility of tannins to invertebrate wood borers which have to eat the wood to digest out the simpler substances.

The Flavonoids

This is a group of complex phenolic derivatives, mostly pigments, which are found commonly in plants but whose role is often rather obscure. The group does, however, include the anthocyanidins and flavonols which are of considerable importance, through the colour of flowers, in many forms of insect pollination as well as contributing to the colour of fruits. It is less easy to understand the reason for autumnal pigmentation of leaves which is also due to these groups of pigments.

The anthocyanidins. Because of the importance of this group in the colouring of flowers and fruits considerable interest has been devoted to their structure, formation and properties. They are composed essentially of two benzene rings referred to as the A and B rings. Sometimes these combine with various sugar units, when they are then jointly referred to as *glycosides*. The sugars make the pigments more soluble in water; they are thus sometimes classed as the *water-soluble* pigments in contrast with the *ether-soluble* chlorophyll and carotenoid pigments.

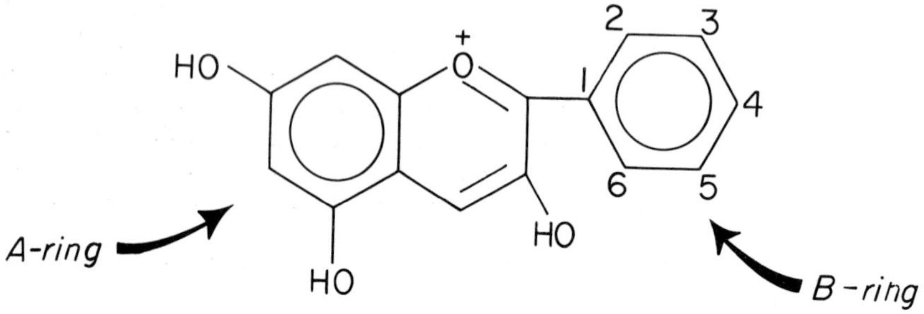

The three simplest anthocyanidins; pelargonidin, cyanidin and delphinidin, have hydroxyl groups substituted into the lateral, B-ring. Pelargonidin, the red pigment named after *Pelargonium* (the garden geranium), has the hydroxyl substituted in the 4-position; Cyanidin, the blue-violet pigment, is one of the commonest anthocyanidins and is found in *Centaurea cyanus* (cornflower) and has hydroxyl groups substituted in the 3 and 4 positions. Delphinidin, the deep-blue pigment that occurs in delphinium and other flowers has substitution in 3, 4 and 5 positions.

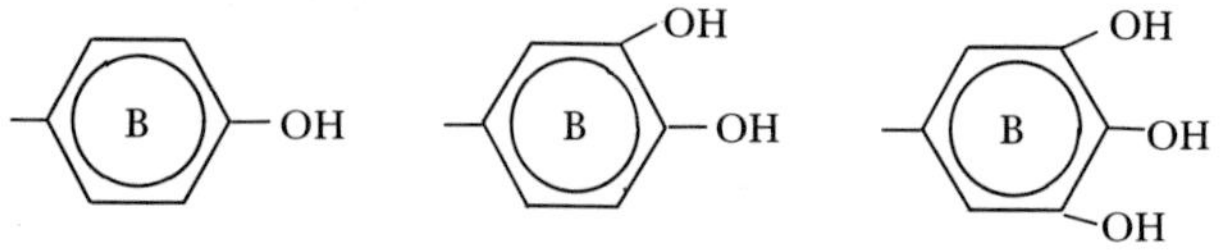

Pelargonidin
(Red)

Cyanidin
(Blue-violet)

Delphinidin
(deep-blue)

Just as an increase in hydroxylation gives added blueness to the pigment, substitution of the hydrogen of the hydroxyl group by a methyl group tends to have the reverse effect, making the colour slightly more red.

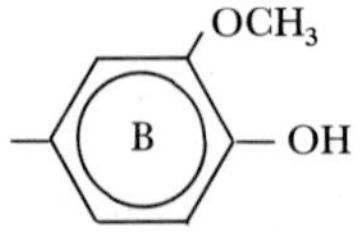

Peonidin
(Crimson-purple)

The final way in which the colour can be changed is by substitution of hydroxyl groups of the main part of the anthocyanidin molecule with sugars. Addition of two such sugar molecules tends to increase the intensity of blueness. For instance, the scarlet pigment pelargonidin becomes a scarlet-magenta.

In most flowers there are mixtures of several pigments; for instance it is usual to find other flavonoids (see below) and different anthocyanidins together. The actual colour of the petals may depend on the relative proportions of these pigments, as well as various physical factors, such as the pH of the cell sap; anthocyanidins are red in acid solution and blue in alkali. Full details for the separation and identification of these pigments are given in the Appendix, p. 246.

Particular interest has been attached to the genetics of flower colours. In most species the presence or absence of a particular hydroxyl group, methyl group or sugar can be related to the effect of a particular gene. In the gloxinia-like greenhouse plant *Streptocarpus* a wide range of flower colours is controlled by four separate gene loci, so as to give a colour range from ivory to deep blue. If all four genes are present only in their recessive states the colour is ivory, no anthocyanidins being formed. When the alleles are replaced by their dominant forms various methyl groups are added or substituted into the B-ring, and at the same time various sugars are substituted into the central part of the molecule:

Colour	*Gene Complement*	*Anthocyanidin*	*Structure*
Ivory	aarroodd	None present	
Salmon	A-rroodd	Form of pelargonidin	
Pink	A-rrooD-	Form of pelargonidin	
Magenta	A-R-ooD-	Peonidin	
Blue	A-R-O-D-	Malvidin	

A number of other gene combinations is also possible. This is a particularly clear example of the way in which a number of genes control the variation in a plant characteristic.

Other flavonoids: Flavonols and Flavones. These are a group closely related chemically to the anthocyanidins and are common water-soluble vacuolar pigments. Unlike the anthocyanidins, with which they are often found, they show little colour range, being almost colourless in neutral or acid solution and yellow or orange in alkaline. They are more common in leaves and photosynthesizing tissues, though their presence is often noticed only in the autumn when the ether-soluble chlorophyll pigments are bleached.

Autumnal coloration

Many plants, such as the copper beech and species of maple, have varieties which possess a large amount of anthocyanidin in their leaves throughout the year. In most plants, however, there is very little or none until the autumn, when the brilliant tints that are so often found are due to extensive changes in the pigment distribution in the leaf. Analysis of the ether-soluble and water-soluble pigments shows that of the former group the chlorophylls are bleached at the beginning of the visible colour change. Later the carotenoid pigments,

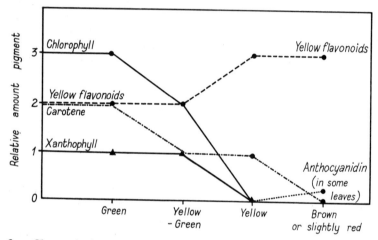

Fig. 6.21. Changes in the leaf pigments during autumnal coloration in the plane tree (*Platanus ×hybrida*).

xanthophyll and carotene disappear. During this time there tends to be an accumulation of flavonoids which results in the leaf taking on a yellow colour, partly due to carotenoids that are still remaining and also to the accumulating flavonoid. During these last stages anthocyanidins may also appear. There is, of course, considerable variation from plant to plant; the pigment changes for the plane tree (*Platanus ×hybrida*) are given in the graph, fig. 6.21. During this colour change, marked changes in the metabolism of the leaf take place. The bleaching of the chlorophyll results in a cessation of photosynthesis, very little starch being formed even in leaves that are brightly illuminated. At the same time the RQ (see p. 72) begins to fall, and it is therefore thought that the formation of anthocyanidins has a high demand on oxygen; the RQ values, determined manometrically, may be as low as 0.5. The reasons for these autumnal colours are more or less obscure; they could be a means for removing waste products of metabolism which might be harmful if translocated into the still living parts of the tree.

Other phenolic substances of importance

Apart from their contribution to the formation of wood and in flower and fruit colour, phenolic substances may be of great importance in other aspects of plant survival. Mention has already been made of the *tannins* which are often bitter tasting and have the capacity of binding with proteins so as to make them less liable to enzymatic destruction. The *cyanogenic glycosides* that are so common in clovers (*Trifolium*) and other leguminous plants may convey similar advantage to the plant through possibly tasting bitter to some herbivores. This can also happen if the plant is damaged by frost, a factor which adversely affects the distribution of cyanogenic strains in colder climates. The glycoside is broken down to release cyanide when damage causes the appropriate enzymes to be released from the plant's own cells. This production of cyanide can be demonstrated quite simply by the use of sodium picrate papers (see Appendix, p. 247).

Protection may also follow an infection; some plants such as pea and bean produce 'warding-off' substances known as *phytoalexins* in response to attack by fungi. Phytoalexins are produced around the fungal lesion and often, though not always, prevent further fungal growth.

Pisatin produced by a phytoalexin the pea

Some of the simpler phenolic substances already mentioned, such as coumaric and related acids, are known to be important in the dormancy of seeds. Seeds that contain such substances will not germinate until rain has leached out the inhibiting acid (see p. 174). Such acids are also known to have *phytotoxic* properties. Leaves of some forest trees, such as the gums (*Eucalyptus*), that are known to contain them inhibit the growth of vegetation in the litter around the parent plant.

6.11 The cell membranes

In discussing these biochemical aspects of physiology mention has been made of the physical problem of keeping the various substances, enzymes and substrates separated from one another. Reactions must only take place at the appropriate time and place; it is the role of the various membranes to ensure that this is so. It is, perhaps, worth examining both the methods by which such separation is achieved and the organelles protected, as well as looking at the possible ways through which membranes may exercise selective control over materials which may pass through them.

Reference to high power electron micrographs (see fig. 6.9) show a number of situations where membranes exist and keep areas of cytoplasm separated. Although the cell wall is an effective separating layer, it is highly permeable to water and many solute molecules, and in any case there is very often cytoplasmic continuity between cells through the cell walls by way of the plasmodesmata (see also figs. 6.20 and 6.23). The term membrane is more usually applied to the outer layer of the cytoplasm, the cell membrane, the plasma membrane or the plasmalemma. It is also applied to the inner cell membrane or tonoplast surrounding the vacuole. These *unit membranes* are just visible as double lines. Other membranes surround the chloroplast, mitochondria and nucleus; these membranes may be more clearly double though in some instances this may be due to a folding of the unit membrane to provide two unit membranes alongside one another. The endoplasmic reticulum often forms a complex internal membrane system stretching almost anywhere in the cytoplasm and also often running from the nucleus right to the cell membrane.

Analysis of the biochemical constituents of the various cell membranes has shown that they are all essentially similar being composed of about half lipid and half protein, though the inner membranes of the various organelles are usually composed of phospholipid and protein. To appreciate the properties of a membrane it is necessary to examine the structure of the individual lipid unit. This is composed of a hydrophilic (water-loving) head and a pair of tails of chains of fatty acids which are water-repellant or *hydrophobic*. The whole molecule is thus described as *amphipathic*.

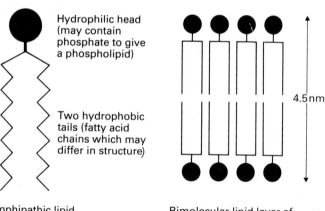

Hydrophilic head
(may contain
phosphate to give
a phospholipid)

Two hydrophobic
tails (fatty acid
chains which may
differ in structure)

4.5nm

The amphipathic lipid
molecule

Bimolecular lipid layer of
the unit membrane

Because of their amphipathic nature the lipid molecules become arranged in a bimolecular layer about 4.5 nm thick with their tails facing in towards one another.

Analysis of the protein constituents of the membrane has been carried out by investigation of their susceptibility to enzyme action, their own enzyme activity and by gel electrophoresis. These lines of investigation have shown that there are three main groups of protein. There are those which lie embedded in the inner or outer lipid layers; these are the *extrinsic* proteins. Other proteins pass through the membrane and often project beyond; these are described as *intrinsic*, and there are those that lie beyond the lipid layers and are often attached to the other groups, binding them together to give strength to the whole system. Some of these possibilities are shown in fig. 6.22.

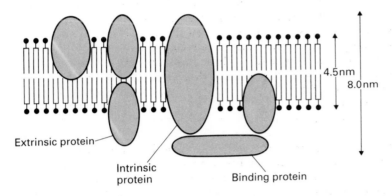

Fig. 6.22. Diagram illustrating the possible arrangement of lipid and protein within the unit membrane of the plasma membrane. (After Singer, S. J. and Nicolson, G. L. *Science* Vol. 175, pp. 720–31, 18 February 1972. Copyright 1972 by the American Association for the Advancement of Science.)

The surface activity of many membranes may be due to the enzymes that are part of the extrinsic group. ATPase is an example of such an enzyme. Others such as cytochrome oxidase are intrinsic. Both must contain hydrophobic amino-acids over parts of their molecules that lie amongst the hydropholic tails of the lipid molecules.

It must be emphasized that what is described above is a highly over-simplified model of the membrane and in life it is not a solid, static arrangement but a mobile, near liquid system and some movement of the constituents is likely, though some of the proteins may bind their surrounding lipids to themselves very tightly.

This kind of unit membrane although only about 8 nm thick has some degree of strength and bounds the cell effectively. How it is able to act selectively in allowing some materials and not others to pass through is more difficult to appreciate. Water-soluble materials may pass through the membrane rather slowly, principally at the pores; lipids may pass through the membrane itself and other substances may be picked up by carriers either at the pores or at the protein zones. Some of the possible mechanisms for the uptake of materials have already been discussed, and it would seem likely that carrier systems, often linked directly or indirectly with respiration, are necessary for the selective absorption of materials, which very often occurs against a concentration gradient. Recently, evidence has been accumulating that some charged ions, in moving in or out of the cell, may carry particular organic molecules along with them, but the activation and control of any such carrier system is still little understood.

The other membranes surrounding the various organelles seem to have very much the same kind of organization though in some situations, particularly in the Golgi apparatus, small vesicles appear to be produced from the individual plate-like *cisternae* that make up the apparatus (see fig. 6.23). These small globules may be *actively secreted* through the cell membrane; a completely different process from the liquid engulfing or 'cell-drinking' called *pinocytosis* that occurs in some plant cells. The connections between the cells, the plasmodesmata, frequently have tube-like structures which are part of the endoplasmic

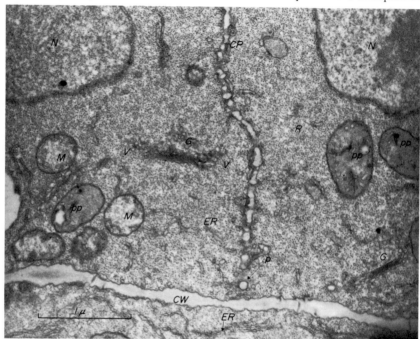

Fig. 6.23. Telophase (late mitosis) in the onion root tip. Note the newly formed cell plate (CP), nuclei (N), older cell wall (CW), mitochondria (M), proplastids pp), Golgi apparatus (G) with vesicles arising from the cisternae at either end (V) and the endoplasmic reticulum (ER). Numerous plasmodesmata (P) are visible in the cell plate and the endoplasmic reticulum tubules run through some of these. (Courtesy of Dr. B. E. Juniper.)

reticulum passing through them and these are probably also important in conveying materials actively from cell to cell.

6.12   Integration of cell processes

In this chapter we have discussed the formation and properties of some of the classes of organic substance that are found in living organisms, the nucleic acids, proteins, carbohydrates, lipids and phenolic substances. Among these biochemical details it must still be realised that the cell works as an organized whole, the nucleic acids of the chromosome form the blue print for ordering the sequence of the various metabolic processes that are going on, by controlling the production of the enzymes and proteins. The process of photosynthesis in the chloroplasts provides the simple sugars that are the chief means for assimilating carbon into the plant and at the same time provide the plant with its energy.

Respiration in the mitochondria releases the energy in the form of ATP, so that it can be used for most of the metabolic processes of the cell. At the same time respiration provides acids which are the starting-point in amino-acid and protein synthesis. Some of these proteins act as enzymes which may enable still further metabolic processes to take place. Yet these, too, are produced to the plan of the DNA of the chromosomes; specific nucleotides of a gene indicating specific proteins or enzymes in the cytoplasm. In some cases the effects of single genes may be apparent in the macro-structure of the plant, for instance, in flower colours, but more often their effects pass unnoticed as part of the vital processes going on in the cell.

Further reading on Biochemistry

BALDWIN, E. (1959). *Dynamic Aspects of Biochemistry.* 3rd edn. Cambridge University Press.
BARKER, G. R. (1968). *Understanding the Chemistry of the Cell.* Institute of Biology's Studies in Biology no. 13. London, Edward Arnold.
CLARK, B. F. C. (1977). *The Genetic Code.* Institute of Biology's Studies in Biology. London, Edward Arnold.
DAVIES, D. D. *et al.* (1964). *Plant Biochemistry.* Oxford, Blackwell Scientific Publications.
HALL, J. L., FLOWERS, T. J. and ROBERTS, R. M. (1974). *Plant Cell Structure and Metabolism.* London, Longman.
HARRISON, D. (1975). *Patterns in Biology.* London, Edward Arnold.
KROGMAN, D. W. (1973). *The Biochemistry of Green Plants.* Englewood Cliffs, N.J., Prentice Hall.
LOCKWOOD, A. P. M. (1977). *The Membranes of Animal Cells.* 2nd edn. Institute of Biology's Studies in Biology no. 27. London, Edward Arnold.
MALCOLM, A. D. B. (1971). *Enzymes.* London, Methuen.
MORRIS, J. G. (1974). *A Biologists' Physical Chemistry.* 2nd edn. London, Edward Arnold.
WALKER, J. R. L. (1975). *The Biology of Plant Phenolics.* Institute of Biology's Studies in Biology no. 54. London, Edward Arnold.
WYNN, C. H. (1973). *The Structure and Function of Enzymes.* Institute of Biology. Studies in Biology no. 42. London, Edward Arnold.
YUDKIN, M. and OFFORD, R. (1971). *A Guide Book to Biochemistry.* London, Cambridge University Press.

7

Coordination in the Plant

7.1   Introduction

In the preceding chapters we have tried to dissect out the various vital processes going on in the plant and have attempted to show how these are co-ordinated within the single cell; in this chapter we have to show how the plant grows and responds as a single organized entity. First, there is the problem of development, involving division and differentiation from a single fertilized egg with a definite gene content, into a complex and specialized adult plant containing many types of cells. In the second place it must be realized that what a plant looks like, and how it behaves internally, are products of the interaction between two groups of factors; the plant's genes, which determine its potential shape, size and metabolism; and also the environment with all its different factors, which may affect and alter the plant considerably, yet within the limits set by its genotype.

7.2   The problem of growth and differentiation

As growth proceeds in any young structure such as the root apex, definite phases of differentiation are clearly visible. In the first phase, involving rapid cell-division, there is considerable manufacture of new protein material, but the increase in size before the next division is not great. In the second phase the beginnings of differentiation are visible, the cells rarely divide and they begin to develop and differentiate into a type of 'adult' cell. This phase may involve considerable elongation and vacuolation, as well as considerable protein synthesis. If the cell is to become one of the vascular elements, for instance a phloem sieve tube, the end walls of the cell become sieve-plates and the cell's own nucleus is broken down. Only a few cells distant the xylem elements differentiate in an entirely different way, eventually having lignified cell walls but no protoplasm. Yet these two sorts of cells were formed from meristematic and elongating cells that were very similar. A similar process, which is still more difficult to explain, occurs in the cells produced by cambium activity. Apparently identical cells produce xylem elements by division to one side and phloem by division to the other. How, then, does this origin of form 'or *morphogenesis* of cells that would appear to be genetically identical take place? Obviously there are two groups of factors involved. First there are the genetic factors, or those concerning the nucleus and the nucleic acids. Second, there is the 'environment' in which the cells are found. The more we can find out about the action and inter-play of these factors, the more we may find out about morphogenesis.

Recently Linderstrøm-Lang has evolved techniques for the chemical analysis of the cell constituents of very small segments of roots. In one experiment of this sort carried out by Jensen *et al.*, the DNA content per cell was analysed continuously back from the root apex. It was found that the DNA content continued to rise in the elongating area, after cell division had more or less ceased (see fig. 7.1). Although it seems unlikely that some cells might have become

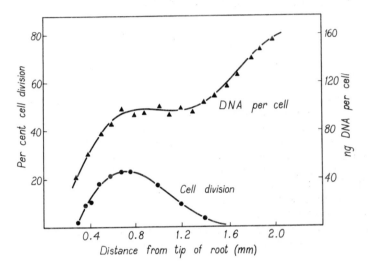

Fig. 7.1. Relationship between DNA content per cell and rate of cell division in an onion root tip. (After Jensen, W. A. (1963). *Plant Biology Today*. London, Macmillan.)

polyploid; that is doubled or increased their chromosome number, nevertheless we may have to discard our concept of genetically identical cells occurring throughout the plant. These histochemical techniques are now being extended to investigate small changes in RNA, protein and enzyme content of the cells.

When a cell is part of a tissue it must be influenced by the surrounding cells in a large number of ways. One of the most important of these is through the action of a wide range of substances referred to as *hormones*.

An example of a secretion produced by one cell which will subsequently influence the development of another is found in the slime-mould or social amoeba, *Dictyostelium discoideum*. This is a remarkable organism that exists as free-living amoebae (see fig. 7.2) that aggregate into a motile 'slug' that then differentiates to form a fruiting body. Although all the amoebae that come together to form an individual slug or grex seem to be genetically identical, the larger tend to become the stalk of the fruiting body and the smaller the spores. The problem is, how are these amoebae organized together into this co-ordinated mass? The organism produces a secretion called *acrasin* which stimulates the amoebae to congregate. They respond not only by movement towards one another but also by producing more acrasin themselves. This will tend to influence more outlying amoebae.

Investigation of the effect of the various factors of the external environment on the growth and development of plants has provided a great deal of information as to how plant hormones control the pattern of growth.

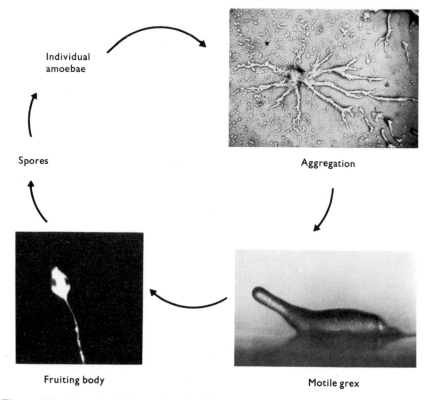

Individual amoebae

Spores

Aggregation

Fruiting body

Motile grex

Fig. 7.2. The life cycle of *Dictyostelium discoideum*.

7.3 The influence of the environment on growth and development

From the moment a seed alights on the ground, it is at the mercy of the environment. It will not germinate unless the suitable oxygen, water and temperature conditions exist, nor will it develop and flower unless it receives sufficient light, water, minerals and nutrients. Yet the environment can only affect the plant within the latter's genetically imposed limits. For instance, the prostrate form of the broom (*Sarothamnus scoparius* ssp. *maritimus*), which grows on exposed cliffs, will remain dwarf and flat-growing under all environmental conditions. However, the scarlet pimpernel (*Anagallis arvensis*) shows considerable *phenotypic variation* and under marsh conditions may become conspicuously succulent (see p. 34).

7.4 The Plant Hormones

It might seem that during the last twenty years there has been great progress in showing how the various environmental factors cause changes in the growth and development of plants through the mediation of a group of hormone-like intermediaries. Unfortunately, although we have now established the presence

Hormone or growth regulator	Formula	Some environmental factors affecting hormone production	Some Processes affected
Auxins	Indolylacetic acid (IAA)	Light Gravity	Phototropism Geotropism
Gibberellins	Gibberellic acid (GA)	Cold temperature Light (via Phytochromes)	Germination Elongation Flowering
Abscisins	Abscisic acid (ABA)	?	Induces dormancy Reduces action of other promotive hormones
Cytokinins	Derivatives of Adenine	?	Promotion of protein synthesis
Ethene		Light (via auxins and phytochromes)	Inhibition of elongation

Table of Plant Hormones. (N.B. The *phytochrome system* is not properly regarded as a hormone and is discussed on page 189.)

of quite a number of such substances and know a great deal about their chemistry in certain specific circumstances, there is still a great deal of difficulty in understanding how they interact with one another. Several hormones may affect a single process and there can be very little certainty as to how the environmental factors are mediated at a fundamental level. In some situations the hormones may be cooperative or synergistic in their action, in others the reverse may be true. One hormone may have similar effects to another and it is often difficult to be sure whether each is acting independently or whether the activity of one is triggering off the activity of the second.

It is clear that most if not all plants coordinate their growth through five main hormone systems. In this discussion the term *hormone* is used for all *naturally occurring* growth regulating substances, whether or not they have an effect on the cell in which they are formed or on a tissue some distance away. The term *growth regulator* is used when dealing with natural and artificial substances in general.

The table on p. 155 lists these main hormone groups, the conditions to which they are sensitive and their main effects within the plant.

7.5   Phototropisms and the Auxins

The bending of plants towards light when grown on a window-sill and the growing of forest trees straighter and faster under a closed canopy are but two examples of *positive phototropism* or the growth of parts of plants towards light. Clearly such an adaptation is of high survival value in competitive situations in enabling the plant to photosynthesize effectively. However, the phenomenon is not confined to green shoots but is found in a variety of structures including the fruiting bodies of many fungi and bryophytes, where it is important that the spores are ejected into the turbulent air layers to be dispersed.

The first serious investigations of tropisms were carried out by Charles Darwin and his son Francis in 1880. They observed that the upper parts of young seedlings of the canary grass (*Phalaris canariensis*) were the first to bend when the seedlings were illuminated from the side. The Darwins then found that decapitated seedlings showed only a poor response to light and also that seedlings, the tips of which had been covered with blackened glass tubes, also usually failed to respond. They concluded that the apex perceived the stimulus and that some influence was transmitted to the lower parts, causing the bending (see fig. 7.3).

Thirty years later, Boysen-Jensen repeated the work of the Darwins using oat coleoptiles. The coleoptile of grasses is particularly suitable for investigations of tropisms; it is a tubular organ (see fig. 7.4) through which the young leaves grow and which protects them from damage in the early stages of germination. Since it is clearly important that the young leaves should be brought into position for effective photosynthesis as early as possible it is not surprising that the coleoptile is strongly positively phototropic. Finally much of the growth of the coleoptile is due to cell elongation rather than cell division, so the mechanics of tropisms can be related to only one growth activity.

In his experiments Boysen-Jensen tried the effect of making horizontal incisions and inserting pieces of impermeable mica on the illuminated or the dark side of the coleoptile. He found that phototropism was inhibited only if the barrier was placed on the *dark* side. If it can be assumed that a growth-promoting

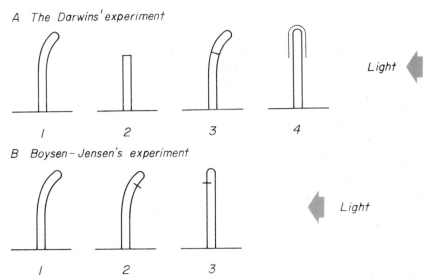

A The Darwins' experiment

1   2   3   4

Light

B Boysen-Jensen's experiment

1   2   3

Light

Fig. 7.3. Response of the coleoptile to light.

- A. The Darwins' Experiment using canary grass seedlings.
    1. Coleoptile responds by bending towards light.
    2. Decapitated coleoptile shows no response.
    3. Decaptitated coleoptile which has had its apex replaced bends towards the light (not carried out by the Darwins).
    4. Coleoptile covered with a black cap does not respond.
- B. Boysen-Jensen's Experiment using oat coleoptiles.
    1. Coleoptile responds by bending towards light.
    2. Coleoptile treated with a piece of mica on the light side still responds.
    3. Coleoptile treated with a piece of mica on the dark side does not respond.

substance is formed evenly in the apex, it would therefore seem likely that some substance, diffusing down the darkened side, was responsible for causing the elongation of the cells on that side and the consequent bending.

The investigations were taken further by Went in 1926. In an effort to determine the effect of the diffusing substance more exactly he placed several cut-off coleoptile tips on an agar block. This was subsequently cut up into a number of smaller blocks and placed, as indicated in fig. 7.5, on one side of the cut stumps of coleoptiles growing in the dark. Bending took place due to the presence of the growth-regulating substance or hormone in the agar. The more tips that had been used the greater was the bending.

There are a number of possible ways in which the bending could be produced. Light could cause the growth-substance or hormone to diffuse to the darker side or light could cause the inhibition, alteration or destruction of the substance, alternatively both situations might occur at the same time. Using biological assay techniques similar to those described above where the angle of bending due to the application of the hormone is measured, the plausibility of these hypotheses can be checked.

The results of such an experiment are illustrated in fig. 7.6. Fresh coleoptile tips were placed as indicated in the diagrams on top of agar blocks. In some

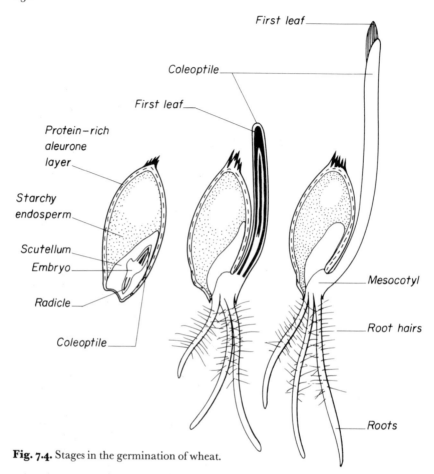

Fig. 7.4. Stages in the germination of wheat.

instances the coleoptile and block was separated by a thin transparent piece of impervious mica. The figures given indicate the relative amounts of hormone in the blocks as determined by the bending test illustrated in fig. 7.5.

From these experiments it is clear that light does not, in this instance, appreciably lower the content of hormone within the whole system by either destruction or inhibition; its main effect is to cause a migration of hormone across the coleoptile to the darker side. This movement has been confirmed by the use of radioactively labelled hormone.

Another important step came in the 1930s with the identification of the hormone, at first called auxin as indolylacetic acid (IAA), a substance which could be synthesized relatively easily in the laboratory.

$$\text{CH}_2\text{--COOH}$$

IAA

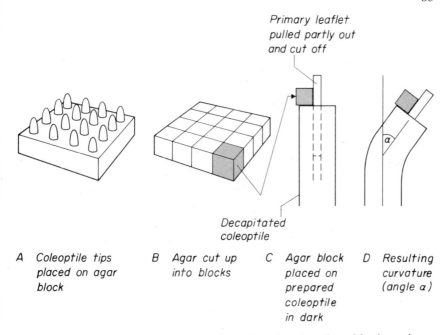

A Coleoptile tips placed on agar block

B Agar cut up into blocks

C Agar block placed on prepared coleoptile in dark

D Resulting curvature (angle α)

Fig. 7.5. Method for isolating growth substances from the coleoptile and for determing the relative amount of substance present.

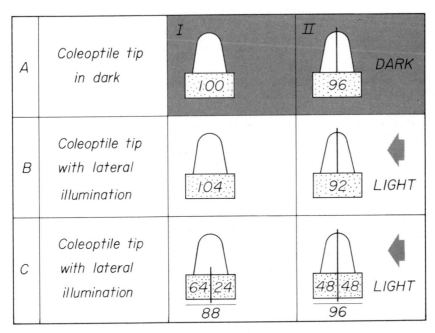

Fig. 7.6. Distribution of hormone as a result of lateral illumination. (Redrawn after Briggs *et. al.* (1959). *Science*, 126, 210–212.)

This enabled other work on the effects of the hormone to be carried out. Work with coleoptiles indicated that the main effect of the IAA was to promote cell elongation, but the investigation of the effect of different concentrations of IAA on the elongation of tissues produced some rather surprising results which have had important agricultural and horticultural consequencies. It was found that the auxin caused cell elongation only at extremely low concentrations. In most stems and coleoptiles elongation is obtained at one part per million and inhibition occurs above ten parts per million (see figs. 7.7 and 7.8, also appendix, p. 248).

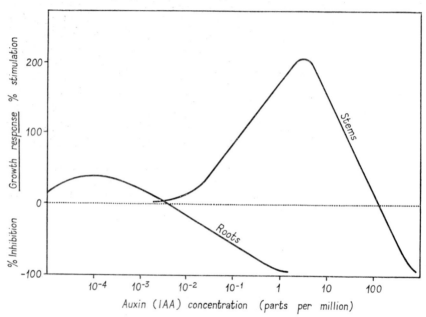

Fig. 7.7. The effect of auxin concentration on growth of root and stem. (After Audus: *Plant Growth Substances.* London: Leonard Hill.)

As different plants have differing tolerances for auxin, the grasses being able to tolerate a higher concentration than the broadleaved plants, this inhibiting property opened up the important new field of selective weed-killers or herbicides. Various substances of similar formula such as indolylbutyric acid have been investigated and are still used but many modern weedkillers, such as 2,4-D(2,4-dichlorophenoxyacetic acid) have a rather different formula.

$$O—CH_2—COOH$$

—Cl

Cl

2,4–D

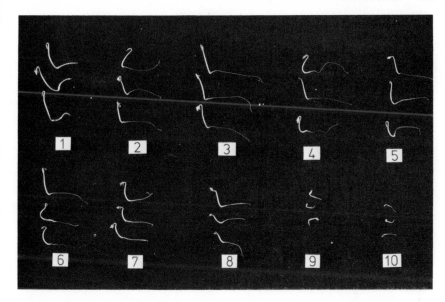

Fig. 7.8. The effect of indolylbutyric acid (IBA) on the growth of cress seedlings.
1 = water, 2 = 0.01 p.p.m, 3 = 0.03 p.p.m, 4 = 0.1 p.p.m, 5 = 0.3 p.p.m, 6 = 1.0 p.p.m,
7 = 3.0 p.p.m, 8 = 10.p p.p.m, 9 = 30.0 p.p.m, 10 = 100.0 p.p.m. (Courtesy Dr. T. A. Hill
and John Marshall, Wye College.)

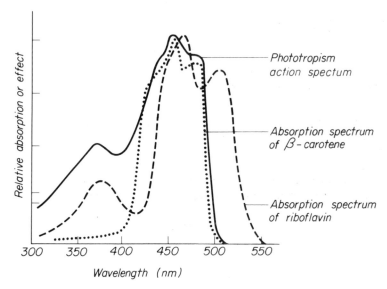

Fig. 7.9. Action spectrum of phototropism compared with absorption spectra. After
Galston in Björkman and Holmgren (1967), *Physiologia Plantarum*, 16, 889–914. Courtesy
of American Scientist.)

Even if it is accepted that IAA is formed in certain tissues such as the apex of the shoot and then translocated to other areas where it may have effects such as elongation, a number of problems still remain. First, in the case of phototropisms, how does light actually cause a particular redistribution of the auxin; second, how is it moved and, third, what is the exact mechanism by which it causes cell elongation or other effects?

The action spectrum for phototropism resembles the absorption spectra of *trans* β-carotene and riboflavin quite closely, both the pigments and phototropism having peaks at about 489 nm. (See fig. 7.9.) It is not yet clear how these pigments pass on their absorbed energy and cause the auxin redistribution.

The means by which the IAA is moved will depend on the plant tissues involved; sometimes it moves from cell to cell whilst at other times it passes through better defined channels such as the phloem.

An effect of IAA that has been investigated exhaustively is that of cell elongation. Addition of auxin to a tissue such as a coleoptile greatly increases the plasticity of the cells (see fig. 7.10). With the decrease in wall pressure the cells

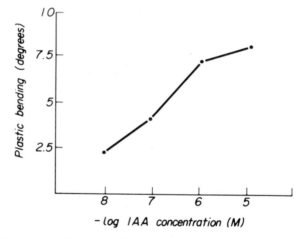

Fig. 7.10. Effect of auxin (IAA) on cell plasticity. (Redrawn after Bonner (1960) *Z. Schweiz Forstr.*, 30, 141–159.)

are likely to have a more negative water potential, water will be taken up, and the size of the cells will increase.

It is also known that if the cell is not respiring normally the change in plasticity will not come about. This would suggest that energy is needed to loosen the cellulose microfibrils. It could be that the energy is needed for the synthesis of enzymes capable of breaking some part of the cell wall constituents, but the effect is a rapid one and so far there is no evidence that the actual cellulose molecules are being broken down.

7.6 Geotropism

Just as plant shoots bend towards light, so roots respond to the influence of gravity and are described as positively geotropic. Stems and flowering shoots are usually negatively geotropic while other structures such as horizontal branches and rhizomes are described as diageotropic. The effect of gravity on the distribution of auxin in a coleoptile has been investigated by similar techniques to that used in the investigation of phototropism. The coleoptile is placed on its side in the dark and the consequent redistribution of auxin investigated as shown in fig. 7.5. Figure 7.11 shows the result of one such experiment.

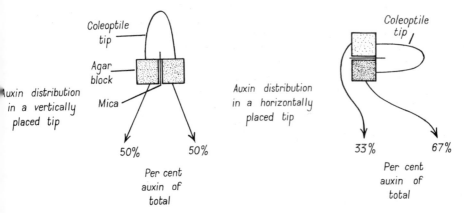

Fig. 7.11. The geotropic response. Auxin redistribution in the coleoptile. (From *Principles of Plant Physiology*, by James Bonner & Arthur W. Galston. San Francisco: W. H. Freeman Co., 1952.)

The inference must be that gravity induces the accumulation of auxin at a high concentration on the lower side of the coleoptile. The cells on this lower side then elongate more rapidly than those on the upper side, causing the coleoptile to curve upwards, away from the stimulus of gravity. It is still not clear what is the *sensor* in geotropism, but a number of cell organelles, such as starch grains (amyloplasts), crystals of various kinds and sometimes the endoplasmic reticulum move under the influence of gravity and may possibly act as *statoliths*. They may help to direct the movement of auxin, though this is still largely speculative.

In the root as opposed to the coleoptile, the situation is rather different. Experiments have shown that it is the extreme apex of the root, normally the *root cap* that initiates the response, possibly by being the site of the sensors or statoliths, but it is thought that the auxin *derives from the shoot* and is merely redistributed through the activity of the sensors. The main difference comes from the fact that the redistributed auxin must reach concentrations on the lower side of a horizontally placed root so that elongation is inhibited, while the cells at the upper surface, relatively deprived of auxin must achieve a concentration suitable for the elongation of cells. Reference to fig. 7.7 shows that the growth of roots is promoted by very low concentrations of auxin while higher concentrations, more like those that would cause elongation in stems, inhibit root growth. The auxin system in horizontally growing *stems* is complex as there must be a delicate balance between the geotropic and phototropic responses. Even more complex are the changes that can occur in a single organ over a period of a few days. For instance, the flower stalk of many species of cyclamen (e.g. *Cyclamen neapolitanum*) grows in a negatively geotropic manner until it has flowered, then it starts to curl up, exhibiting positive geotropism, and eventually more or less buries the young seed head in the ground (where ants, the principal agent of dispersal, are more likely to find the seeds).

Auxins and root formation

The production of roots from residual meristems such as the cambium and pericycle can often, but not always be greatly increased by local application of auxin. This suggests other important natural effects of the auxins but the property has been put to use in the rooting of cuttings in horticulture. Indolylbutyric acid (IBA) and naphthaleneacetic acid (NAA) are often used as

Tall plant,   Dwarf plant,   Dwarf plant,
untreated   treated with 10.24 μg   untreated
              gibberellic acid on
              the first true leaf

Fig. 7.12. The effect of gibberellic acid on the growth of genetically dwarf strain of pea (*Meteor*).

these substances are less readily translocated and therefore more local in their action than IAA. Preparations of the hormone in an inert powder form are often used. The cut stem is dipped first into water and then into the cutting powder. Rooting is greatly enhanced and the technique has found considerable commercial application in horticulture.

7.7   The gibberellins

A number of parasitic diseases of plants cause considerable changes in the growth form and fertility of the host plant. Few are more dramatic than the effect of infection of rice by the fungus *Gibberella fujikuroi*. Many years ago the Japanese noticed that the infection caused a rapid increase in height of the host plant though it was rendered spindly and sterile. However, it was not until the 1950s that the main active substance, gibberellic acid (GA) was finally isolated and the general importance of the gibberellins as powerful growth-promoting substances in most higher plants finally realized.

One of the first important observations regarding the action of GA was that not all plants are equally sensitive to its application. In maize and pea for instance, some genetically dwarf strains are highly sensitive while tall strains are hardly affected. Figure 7.12 shows the effect of GA on the growth of a dwarf strain of the garden pea.

Fig. 7.13. Sterile halves of barley seeds without embryos, treated with gibberellic acid (GA). *Left*, treated with water only; *Centre*, GA (1 part/10^{12}); *Right*, GA (100 parts/10^{12}). The right hand seed shows that the starch endosperm has been digested. (Courtesy Dr. J. E. Varner, Michigan State University.)

At first sight it might appear that the dwarfs simply lacked the gene necessary for initiating the production of GA, but the situation is not so simple as it might seem since assay has shown that not all tall plants contain more GA than their dwarf counterparts (see appendix p. 249)

Another important effect of GA concerns the germination of seeds. In cereals GA is produced by the moistened embryo and this triggers off the synthesis of the starch hydrolysing enzyme amylase in the aleurone layer (see also fig. 7.4) surrounding the endosperm. If barley grains are surface sterilized and then cut in half to remove the embryo (see appendix page 249) the remaining half can be used as a method for assaying the presence of GA by the amount of starch digestion and sugar formation that takes place in a given time. This effect is illustrated in fig. 7.13 above. In the intact seed the embryo would be the source of the natural GA (see appendix p. 249).

The action of GA on barley seeds has been shown to be due to the effect of the hormone on increasing the production of the messenger RNA, responsible for the formation of the amylase, from the DNA of the nucleus. This indicates that the important action of the hormone may be to 'switch on' a particular gene which has previously been suppressed by some other factor.

This work may help to make sense of the many and varied properties that have been ascribed to GA. The most obvious effect of the hormone, that of causing the rapid elongation of stems may be brought about by the GA allowing for or at least enhancing the production of IAA. This could account for the *synergistic* or apparently co-operative action of these two hormones, illustrated in fig. 7 14.

Many of the other activities of GA are related to several processes associated

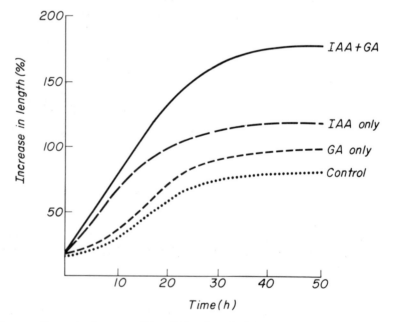

Fig. 7.14. Synergism between gibberellic acid and indolylacetic acid shown in the elongation of pea stem sections. (After Ockerese, R. and Galston, A. W., 1967, *Plant Physiology* 42; 47–54.)

with reproduction including vernalization, flower induction and fruit formation; these situations are discussed in Chapter 8.

7.8 Abscisic acid

Many of the hormones we have so far discussed have growth-promoting effects, however for any plant to survive difficult environmental conditions it must be either vegetatively dormant or have dormant seeds. Wareing, working on birch trees in 1964, found increasing quantities of a dormancy-inducing substance in the leaves during the latter part of the year. When this was eventually isolated it was found to be an isoprenoid (see p. 142) and it was subsequently named abscisic acid (ABA). When this substance is supplied to growing shoots of woody plants a number of morphological changes take place; small scale-like protective leaves are formed, growth decreases and leaf fall or abscission may follow. Growth of dormant buds and seeds at the right time of year or under appropriate conditions is also inhibited. The production of ABA in these cases is probably linked to the day length; the shorter days of autumn promoting ABA formation.

The action of ABA seems to be in opposition to the growth-promoting GA. This has been shown by its effect on the release of amylase that GA can initiate from the aleurone layer of barley seeds (see p. 165). The graph in fig. 7.15 shows

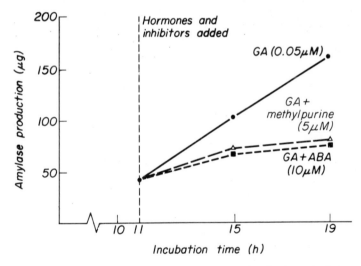

Fig. 7.15. Inhibition of Gibberellin induced amylase production in excised aleurone layers of cereal seed by abscisic acid. (After Chrispeels, M. J. and Varner, J. E., 1967, *Plant Physiology* 42; 1008–1016.)

how addition of ABA depresses amylase production. It also shows that a similar inhibition can be achieved by addition of a known inhibitor of RNA synthesis, *methylpurine*. This suggests that ABA has a directly opposite effect to GA in that it inhibits rather than promotes RNA formation.

7.9   The Cytokinins

This group of hormones also owes its discovery to what at first sight appears to be a rather unusual choice of material in certain growth investigations—coconut milk. In 1941 van Overbeek was investigating the conditions needed to allow very young embryos to continue growth in culture media. Coconut milk, which is liquid endosperm containing many free-floating nuclei, was found to have strong growth-promoting properties. Although IAA can cause the onset of cell division, particularly in the residual meristems of stem and root (hence the use of auxins like indolylbutyric acid, IBA, as inducers of the rooting of cuttings in horticulture), coconut milk is a much more vigorous inducer of mitosis in most tissues. This might suggest that it contains some precursor or material involved in the activity of the nucleus. Skoog and others then identified a range of substances that we now call *cytokinins* that are capable of stimulating mitosis as well as a number of other physiological effects. These are all chemically related to the purine *adenine*, a constituent of nucleic acids. *Kinetin* (furfuryl adenine) is a synthetic cytokinin that is quite easily manufactured from adenosine and has been widely used in growth experiments.

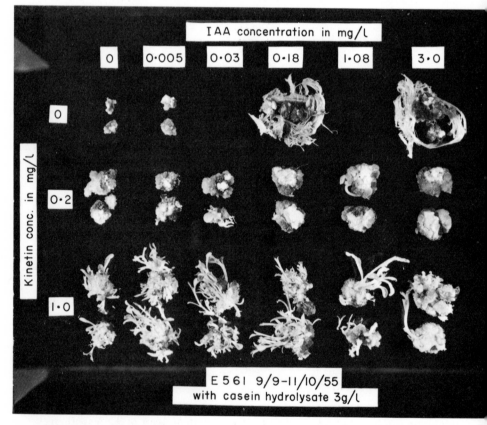

Fig. 7.16. Interaction between Kinetin and IAA on tobacco culture. (Skoog I., and Miller, C. O., (1957) *Symp. Soc. Exp. Biol. 11*: 118–131.)

In 1957 Skoog found that sterile tobacco tissue, supplied with nutrient containing both kinetin and IAA in culture would recommence cell division, and differentiation into root or shoot would take place. It was of considerable interest that IAA on its own only allowed for root formation while kinetin, particularly at high concentrations tended to produce shoots. Intermediate concentrations of both produced only *callus* growth (undifferentiated masses of cells). This undifferentiated growth resembles a gall and is produced by rapid mitosis without cell specialization. Similar growths are due to a number of plant infections; many 'witches brooms' which are caused principally by bacteria, produce potent cytokinins. These interactions between IAA and the cytokinins are shown in fig. 7.16. Here again then we have another example of two plant hormones interacting in the way that they control development.

Work of this kind is central in the study of developmental plant physiology and the possibilities of being able to refine such techniques to produce whole plants from tissue culture are of great potential value. The culture of single cells or small groups of cells, usually derived from apical meristems, for the production of disease-free whole plants is now standard practice in horticulture. Recent work has shown that it is possible to grow pollen in culture media not only to produce pollen tubes but also to allow these to grow, produce cells and differentiate to produce whole haploid plants. The haploid plant can then be converted into a homozygous diploid by the use of the alkaloid colchicine. This will make the selection of homozygous strains showing characteristics that may be useful for the plant breeder much quicker and easier.

Recent work has shown that cytokinins may be found on transfer RNA molecules, their site being adjacent to the point of attachment between the transfer and ribosomal RNA. In this way they may exert a profound effect on the capacity of the cell to synthesize proteins.

Just as cytokinins seem important in allowing for cell growth and maybe some aspects of differentiation, so they also have an interesting property in helping to maintain organization and therefore delay senescence. A rather similar property involves the dormancy of lateral buds; local application of kinetin may allow these to loose their dormancy and begin growth. If this happens under natural conditions it may well result in a profound change in the overall shape of the plant.

In horticulture and agriculture there has recently been the development of a synthetic hormone referred to as *cycocel* (CCC) which has some properties resembling the cytokinins, though the formula is quite different. CCC causes a spreading of the apex and an increased branching of the plant. This has proved of great value in the growing of chrysanthemums where it is desirable to produce short and bushy plants. The substance is also being used to reduce the height of some cereal crops and to prevent or reduce the beating down or 'lodging'; in this it is thought to behave more as an inhibitor of gibberellin synthesis.

7.10 Ethene (Ethylene)

The organic gas ethene, C_2H_4, might seem a curious kind of hormone or at least growth regulating substance, but its effects on plant growth are as clear and probably as important as the other hormones. In early work ethene production was thought to be associated with the decay of fruits, particularly of apples, but it is now known to be produced naturally in a great many plants and parts of plants, and not only as a result of infection. Development of gas chromatography has made it possible to detect very small quantities of ethene and as a result it has been implicated in quite a number of physiological phenomena. If ethene is applied to pea seedlings in the dark, the growth form of the seedlings is greatly altered; elongation is inhibited, radial growth stimulated and the curved apical hook of the developing leaves unfolds (see fig. 7.17). The response of the plant to gravity is also greatly reduced.

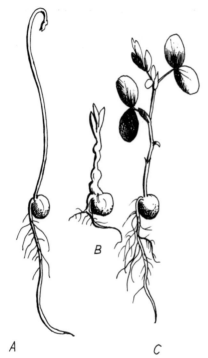

Fig. 7.17. Effect of ethene on pea seedlings. Seeds germinated 20 days. (A) In the dark, no ethene treatment. (B) In the dark, ethene treated. (C) In the light, no ethene treatment. (Class project result.)

Ethene is also produced in flowers and fruits. In some orchids pollination is followed by a rise in ethene production which may help bring on the senescence of the flower and the onset of fruiting. The rise in respiratory rates associated with the *climacteric*, or full ripening, of the fruit is also associated with ethene formation. Various hormones such as IAA may stimulate ethene formation and the gas itself is thought to have effects on membrane permeability and enzyme formation, so it may be that ethene acts as a final step in the sequence of events between stimulation and developmental effect.

Recently the ripening of fruits has been greatly accelerated by the use of an aerosol spray containing an artificial ripening agent that yields ethene; 2-chloroethylphosphoric acid ('ethrel'), sprayed on to tomatoes causes the rapid ripening of the fruit. This is of considerable economic value in clearing greenhouses of a late tomato crop.

Further reading on Coordination in the Plant

ASHWORTH, J. M. and DEE, J. (1975). *The Biology of Slime Moulds.* Institute of Biology's Studies in Biology no. 56. London, Edward Arnold.

AUDUS, L. J. (1972). *Plant Growth Substances.* London, Leonard Hill.

BLACK, M. and EDELMAN, J. (1970). *Plant Growth.* London, Heinemann.

GALSTON, A. W. and DAVIES, P. J. (1970). *Control Mechanisms in Plant Development.* Engelwood Cliffs, Prentice-Hall.

GEMMELL, A. R. (1969). *Developmental Plant Anatomy.* Institute of Biology's Studies in Biology no. 15. London, Edward Arnold.

HILL, T. A. (1973). *Endogenous Plant Growth Substances.* Institute of Biology's Studies in Biology no. 40. London, Edward Arnold.

MAYER, A. M. and POLJAKOFF-MAYBER, A. (1975). *The Germination of Seeds,* 2nd edn. Oxford, Pergamon.

SALISBURY, F. S. (1963). *The Flowering Process.* Oxford, Pergamon.

WAREING, P. J. and PHILLIPS, I. D. J. (1970). *The Control of Growth and Differentiation in Plants.* Oxford, Pergamon.

8

Physiological Organization and the Life Cycle

8.1 Introduction

Photosynthesis, stomatal movement and sleep movements of leaves and flower petals are just some examples of the diurnal or circadian rhythms that go on in a plant from day to day, but the plant must also be able to adjust its whole life to the changing seasons. If the plant is an annual or ephemeral it must be able to germinate and grow quickly once conditions are favourable; equally it must be able to produce seeds capable of withstanding unfavourable conditions for considerable lengths of time. In perennial plants the degree of synchronization and adjustment to the seasons that is necessary may be even greater. Germination of seeds and the sprouting of leaves must usually occur in early spring, but not so early that frost may kill the young leaves. Photosynthesis and growth should be optimal during the summer and flowering must take place sufficiently early for seeds and fruits to be formed before the less favourable conditions of autumn and winter set in. The physiological activity of the growing shoots must also be modified and reduced so that leaf fall, if it takes place may do so with the first serious autumn gales and frost. In many of these instances the plant may have to *anticipate* the onset of the unfavourable conditions; otherwise it may be placed in an uneconomic and even vulnerable position.

Every plant and each environmental condition may be associated with its own special problems; for instance in many cases it is not the winter but the summer which may be the difficult season. Around the Mediterranean and in similar climatic areas such as South Africa and California, much growth takes place in the winter and early spring and the plant may be more or less dormant in the hot dry season (see discussion of xerophytes on page 36).

This chapter sets out to examine the various phases in the life cycle of higher plants and endeavours to show how some of these stages are synchronized with the seasons through a variety of control systems.

8.2 Control systems

This synchronization will be affected in part at least, by one or more external or exogenous factors such as light, temperature and rainfall. Of these the most dependable seasonally is the daily duration of light, but temperature, through its direct effects on growth rates as well as the more particular effects of frost, is also very important. Finally, seasonal rainfall will be vital in making it possible for any growth to take place at all.

These 'setting mechanisms' that synchronize the growth of the plant with its particular environment must have appropriate 'sensors' or receptive systems within the plant if the plant is to respond. Light affects a number of pigments other than those concerned with photosynthesis and some of the hormone systems discussed in the previous chapter are influenced by light. Blue light absorbed by the carotenoids and riboflavin causes a change in the level of IAA and this influences the shape and spread of the plant. Red light absorbed by the *phytochrome system* (see p. 180) is important in controlling germination and flowering, while cold may alter the balance of gibberellins, increased concentration of which may also assist germination and flowering.

Alongside these hormonal effects the three environmental factors have equally important effects on general metabolism and so again help to control the overall size and shape of plants. However, whether or not a plant responds to a particular set of environmental conditions depends not only on its age and physiological state but also on its genetic character. Quite closely related plants such as strains of annual and biennial henbane (*Hyoscyamus*) may only differ from one another in respect of one gene yet their responsiveness to the environment and the whole pattern of their life cycles is very different.

8.3   Seed dormancy

Once seeds have been dispersed from the parent plant they may not germinate for some time, even though the environmental conditions appear to be favourable; a period of *after-ripening* is necessary. This internal dormancy may be due to a number of factors; first the embryo may be incomplete in its development, second there may be inhibitory substances present in the seed or fruit, and finally the seed coat or *testa* (see fig. 8.1) may be impervious to water or other vital external requirements for germination.

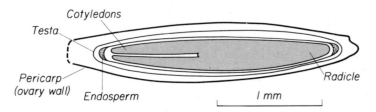

Fig. 8.1. Transverse section of lettuce seed. (Contrast wheat shown in Fig. 7.4 on p. 158.)

Incomplete development is quite a common feature in flowering plants but is most frequent in seeds of parasitic, saprophytic and symbiotic plants such as members of the orchid family and broomrapes (*Orobanchaceae*). It is also common in the buttercup family, occurring in the lesser celandine (*Ranunculus ficaria*). In these the embryo is anatomically incomplete but in other cases, such as juniper, the embryo appears completely formed but certain physiological changes have still to go on. After-ripening may occur in some species when the seed is dry stored, or, alternatively when it is wet and in the *imbibed* state. The latter situation involves a process referred to as *stratification* and has to be used in some commercial situations such as the growing of apples from seed where it is customary to bury the whole apple for some months in moist sand before removing the pips and sowing them in the normal way.

Inhibition of seed germination may, in a sense, be caused more by external than internal factors, as the fruit itself may produce substances that prevent germination. The most likely cause of this is that the high concentration of sugars in the fruit produce osmotic inhibition. The situation can be illustrated by treating seeds with solutions of sodium chloride. Figure 8.2 illustrates the effect of this on the germination of alfalfa or lucerne (*Medicago sativa*) and also on lettuce. Alfalfa is capable of living under near desert conditions with soil of relatively high salinity and, as would be expected, is less inhibited by the addition of the sodium chloride to the culture medium.

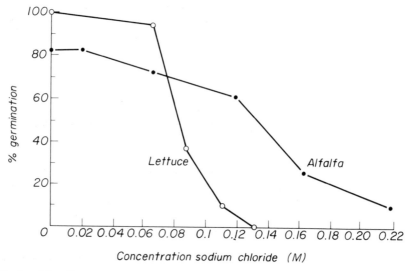

Fig. 8.2. The effect of sodium chloride on the germination of seeds of lettuce and alfalfa. (After Mayer, A. M. and Poljakoff-Mayber, A., *The Germination of seeds*, Pergamon 1963.)

More complex inhibitors, found in various parts of the fruit and seed may also inhibit germination. Coumarin, caffeic acid, cinnamic acid, ferulic acid and benzoic acid have all been implicated in this inhibition and it is generally supposed that germination is prevented until the concentration of the inhibitors is lowered through leaching by rain and water (see appendix p. 251). Identification of natural germination inhibitors has led to the investigation of natural germination stimulants. Apart from the hormonal stimulants such as gibberellic acid (GA) and the cytokinins, ethene, thiourea and potassium nitrite have all been shown to stimulate germination at certain concentrations. All these substances have been identified in some living plants. We are a long way from understanding the detailed mode of action of these diverse substances that have so profound an effect on germination, but the use of artificial germination inhibitors such as 'propachlor' has proved to be of horticultural and economic value.

The third way in which germination may be prevented is through the condition of the seed coat. The testa may be impervious to water and air until it has softened and the mechanical constriction of the contents of the seed may also prevent its germination. The cocklebur (*Xanthium*) has been extensively investigated with regard to the permeability of the seed coat to oxygen. Its fruit contains two seeds, one above the other; the top seed will only germinate in high oxygen concentrations but the lower will grow in comparatively low concentrations. That this is due to the permeability of the seed coat has been shown by removing the testas carefully and allowing the excised embryos to germinate on their own. Very low concentrations of oxygen are then sufficient for the germination of both kinds of seed. This is shown in the table below. This difference in germination requirement is presumably of ecological significance in enabling seeds to germinate at different times or under different conditions.

| | *Oxygen (%) required for 100% germination of seeds or embryos* | |
	Intact seeds	Excised embryos
Upper seeds	100	1.5
Lower seeds	6	0.6

Effect of oxygen on the germination of the seeds of cocklebur (*Xanthium*) (Data from Thornton, N. C. (1935) Contr. Boyce Thompson Inst. 7477).

The ways in which this form of dormancy may be broken are largely external and through the activity of agencies in the soil. This may take place through abrasion and the activity of the humus acids as well as micro-organisms such as bacteria and fungi. The small soil invertebrates may also be of importance here. Passage of seeds through the digestive tracts of larger animals also often increases the permeability of the seed coat. Most gardeners know that hard seeds, particularly members of the *Leguminosae*, such as lupin, germinate much more rapidly if the testa is nicked with a knife before the seed is sown.

All these dormancy adaptations may result in both different individuals as well as different species having very different germination rates, which will be of considerable survival value to the plant in the wild. It seems probable that natural selection has favoured systems producing staggered germination and there are examples of genetic systems which allow for this to happen. In the shoo-fly plant (*Nicandra physaloides*) the diploid state normally consists of 18 chromosomes together with a pair of iso-chromosomes. At meiosis, one of the iso-chromosomes is liable to be lost; in consequence populations of *Nicandra* occur with both 19 and 20 chromosomes. Seeds with twenty chromosomes germinate first, but those with the smaller number may germinate years later. Many weed species have mechanisms which give periods of dormancy which are highly variable but may be very long. On the other hand, in the domestic selection of crop plants, genetic differences allowing for uneven germination will be eliminated quite rapidly so that the crop grows and matures as evenly as possible.

Longevity of seeds

While these internal factors may result in a delay of germination for some time, unsuitable external conditions are more likely to prevent germination over longer periods. Many examples of the longevity of seeds have been quoted, but the record is probably a lupin from Alaska the seed of which retained its viability for about 10000 years through being preserved under extremely cold conditions. Other examples include the lotus (*Nelumbo nucifera*) a relative of the water-lily, the seeds of which were discovered in a peat bog which could be dated using ^{14}C. Such seeds many hundreds of years old have retained their viability. More common plants such as the grain crops show a steady decline in viability on storage so that few seeds older than ten years germinate. Many garden weeds on the other hand such as the shepherd's purse (*Capsella bursa-pastoris*) may retain their viability for as long as twenty-five years; in the meantime bouts of germination may occur when both environmental conditions and internal situations are favourable (see fig. 8.3).

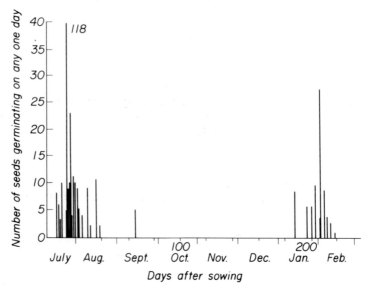

Fig. 8.3. Intermittent germination of seeds of the Shepherd's purse (*Capsella bursa-pastoris*): 63 percent germinated in the eight months shown. The remainder might germinate over the following twenty-five or more years. The seeds were sown on ordinary garden soil and show germination peaks in autumn and spring; it would seem likely that this staggered germination is due, at least in part, to internal control of dormancy. (Data of Salisbury, Sir E., from *Biology of Garden Weeds*, Journal of the Royal Horticultural Society 87, 9.)

8.4 Germination

Provided the seed is ripe and ready to germinate, the main conditions required are water, a suitable temperature, oxygen and either light or dark depending on the seed in question. The germinating seed must have the right balance of environmental conditions; very often the seed is shed under relatively dry and warm conditions, but there are many instances where seeds are dispersed on to moist soil but do not germinate because the temperature is unsuitable. This has been shown for annual plants growing in the Colorado desert; when the seeds were sown on moist soil at low temperatures around 10°C. only winter annuals germinated, while summer annuals (including more xerophytic and less hardy species) only germinated at much higher temperatures in the 26–30°C. range. Interaction of these two factors with the oxygen in the soil atmosphere is also important and has been demonstrated with the lesser reedmace or bullrush (*Typha angustifolia*). In this instance germination, which also usually requires light, is better at higher temperatures and at *low* oxygen tensions. This is illustrated in the table below.

Temperature %C.	% Germination	
	20% oxygen	2% oxygen
15	20	0
20	37	0
30	61	96
35	48	0

Germination of seeds of *Typha* in the light on moist blotting paper at varying temperatures and oxygen tensions. (Data of Sifton (1959) from Mayer, A. M. and Poljakoff-Mayber, A. The Germination of Seeds, Pergamon 1963.)

Water and Germination

The uptake of water into the seed is the first important event in germination. This may take place through the micropyle or almost any point in the seed coat.

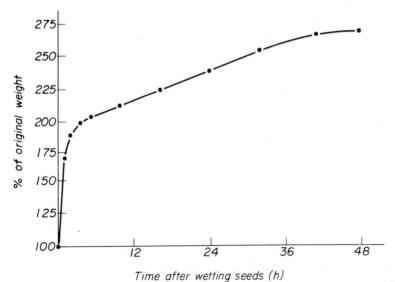

Fig. 8.4. Imbibition of water by lettuce seeds. (After Levari, from Mayer, A. M. and Poljakoff-Mayber, A. *The Germination of Seeds*, Pergamon 1963.)

The dry seed is packed with colloidal materials with strong water imbibing properties and, as metabolic activity begins, simpler substances such as sugars will be present; the combined effects of these is to give the seed a strongly negative water potential, so that water is usually taken up very rapidly; this is shown in fig. 8.4.

The hydration of the colloidal particles, particularly the proteins, and the expansion of the cells of the embryo will result in a considerable pressure being exerted on the fruit wall or the testa. In some cases, where the fruit is hard, as in nuts, this may produce a pressure as high as $350\,kNm^{-2}$ on the fruit wall which is usually sufficient to rupture it in its already softened state.

The hydration of the cells of the embryo and other parts of the seed will allow for the activation of enzymes of which three main groups, acting on the main food storage substances, are likely to be important. The amylases will act on starch, lipases on fats and proteases on protein food reserves. This is illustrated in fig. 8.5 which shows how these enzymes act in seeds with different food reserve substances. The various sugars, fatty acids and amino-acids released through the activity of these enzymes will allow for a rapid increase in the rate of respiration and the general processes of metabolism, cell division and growth will follow. However these activities are likely to be particularly influenced by the temperature and oxygen tension of the soil.

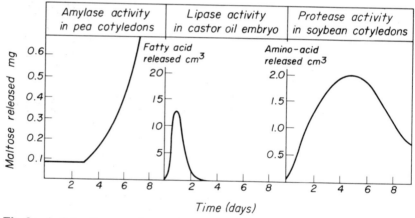

Fig. 8.5. Activity of enzymes in germinating seeds

1. Fatty acid released. This is expressed as a titer with 0.02 M Sodium hydroxide.
2. Amino acid released. This is expressed as a titer with 0.025 M Potassium hydroxide.

(All diagrams modified after Mayer, A. M. and Poljakoff-Mayber, A. *The Germination of seeds.* Pergamon 1963. Lipase activity from data of Yamada, M. (1957), Sci. Papers Coll. Gen. Ed. Univ. Tokyo, **7**, 97.)

Temperature and germination

Although an increase in temperature would be expected to speed up most metabolic processes, both the actual occurrence of germination and the rate of the process are not always directly related to temperature. Most seeds germinate over a standard range of temperatures which are related to the environment in which the plant grows. The following table shows this for a number of species found in different climates.

Species		Temperature °C.		
		Minimum	Optimum	Maximum
Rice	*Oryza sativa*	10–12	30–37	40–42
Maize or corn	*Zea mays*	8–10	32–35	40–44
Wheat	*Triticum sativum*	3– 5	15–31	30–44
Barley	*Hordeum sativum*	3– 5	19–27	30–40

Data from Mayer and Poljakoff-Mayber.

Provided that seeds are germinated within their appropriate temperature range, germination is usually quicker at the higher temperatures, though interaction with other factors such as oxygen and light may also be important (see table on p. 177). Very high temperatures, induced by burning, may actually be necessary for the rupture of fruits so that germination can take place in some plants living in near-tropical forests and scrubland. In Australia, the honeysuckle trees (*Banksia*) show a prodigious germination after a bush fire, rapidly covering the burned areas between their larger parents. Hitherto competition from other scrub plants would have made their survival impossible.

Frost and the cold-treatment of seeds are environmental conditions that are possibly more complex in their effects. Many seeds germinate much better if they receive cold treatment prior to being in their normal growth temperature such as given in the table above. The cold treatment is usually only effective if applied to the seeds when they are in the wet, imbibed state, that is, during stratification. But it is still not clear exactly what physiological changes are involved though the hormone balance, particularly of the GA relative to other hormones, is probably involved so that inhibiting substances are reduced or counteracted. Cold treatment or fluctuating diurnal cold/warm treatments are necessary for the germination of a wide variety of seeds, particularly members of the rose family (*Rosaceae*), buttercup family (*Ranunculaceae*) and iris family (*Iridaceae*). The capacity of the plant to 'sense' this probable winter cold may provide an important mechanism for triggering spring growth and development. In some plants cold-treatment of the seed both before and during germination may result in the plant completing its life cycle much quicker. The term *vernalization* used to be used to describe such treatment of the seed during the very early stages of germination, usually before there are any outward and visible changes so that the plant may come into flower much earlier. At the present time the term is more often used to describe the cold conditions of winter that may be needed to induce flowering in biennials and other plants.

The above effects are brought about by temperatures above freezing but frost may have other effects. Many seeds, due to their highly dehydrated state are unharmed by quite low temperatures. It is also possible that frost, by causing the formation of ice crystals, could accelerate the breakdown of the seed coat. Once the seed has started germination and growth the chances that frost damage will be fatal increases greatly; most plants possess a high concentration of dissolved substances in their cell sap but, should ice crystals form within the cell, death of that cell is almost inevitable due to rupture of the cell membranes and dehydration of parts of the cytoplasm. In this way frost must be a powerful selective agent in influencing the growth and development of plants and in limiting their geographic distribution.

Light, dark and germination

Light and dark seem extremely variable in their effects on germination. Some plants living in fairly wet conditions such as purple loosestrife (*Lythrum salicaria*) and hairy willow-herb (*Epilobium hirsutum*) require light while the garden annual love-in-a-mist (*Nigella*) germinates best in the dark. On the other hand most cultivated plants and crops are indifferent to light and dark treatments with regard to germination.

In other cases the effects of different day lengths and different qualities of light have been shown to be very important; here germination may closely resemble flowering in its dependence on the day length (photoperiod). Lettuce seeds have been fully investigated in this respect. It is first necessary for the seed to be in its imbibed state and they should be soaked in water for a few minutes at their optimum temperature for germination (about 25°C.) before any particular light treatment is investigated. The seeds become sensitive to light quickly on wetting so it is necessary to ensure that the timing of the procedure is carefully controlled.

One interesting feature of lettuce seeds is that they respond differently to light, this is shown in the table below (see also appendix p. 250):

| | Variety of Lettuce | | |
Treatment	Grand Rapids	Cannington Forcing	Great Lakes
Dark	10	5	86
Light	94	90	86

Percentage germination of lettuce seeds of different varieties in light or dark after an initial imbibition period. From the Teachers' Guide Volume 2 Nuffield Advanced Biological Science page 89.

With Grand Rapids it is clear that white light promotes germination. In an effort to determine the quality of light responsible for this promotion it was found that red light of wavelength 660 nm also promoted germination while far-red light of 730 nm wavelength was even more inhibitory than dark treatment. This situation, which is similar to flower induction (see p. 189) is a reversible situation, the final treatment of light or dark, red or far-red light being the vital one in determining whether or not germination will take place (see table opposite).

Difficulty is sometimes encountered in repeating such experiments and it is known that the light requirement may change and decrease with the age of the seed. It is also possible that genetic changes have taken place due to selection by commercial growers so that the strains marketed in one part of the world may differ in their responsiveness from others of similar name. Other plants such as love-in-a-mist (*Nigella*) show rather similar responses and can be used instead of lettuce.

The ease of reversibility of the situation with different qualities of red light suggests that some pigment may be involved which changes its state according to the light stimulus. Determination of the action spectrum of germination has suggested that the blue-green pigment *phytochrome*, which is also involved in flower-induction, is the substance concerned in the perception of the stimulus. It is not yet clear how the accumulation of the appropriate form of phytochrome stimulates the seed to begin growth but the effect of GA is particularly interesting in that it seems most effective in reversing far-red inhibition of germination (see table below). In other words it could be that as the appropriate form of phytochrome accumulates GA is formed and this hormone, through allowing for the increased activity of the hydrolytic enzymes in the embryo and food reserves, as well as helping to promote elongation, greatly accelerates the process of germination.

Treatment	Percentage germination	
	In water	In GA 2.9×10^{-5} M
Dark	12	39
Red	44	67
Far red	5	25
Red then far red	11	35

Effect of light and GA on the germination of lettuce seeds at 26°C. (After Evenari *et al.*, after Mayer, A. M., and Poljakoff-Mayber, A., page 95.)

There may be commercial and horticultural applications here for some seeds, particularly those of high alpines which are notoriously difficult to germinate, as treatment with GA at 1000 p.p.m. for 24 hours has produced remarkable improvement in germination rates.

Finally it must be emphasized that seeds will only germinate successfully under natural conditions if both the seed is ripe and the correct balance of external conditions prevails. A sudden change in the latter may spell death to the plant if germination has already begun. At this time the plant is perhaps at its most vulnerable stage, imperfectly established in the soil and as yet unable to exercise both proper water uptake and effective photosynthesis. For these reasons this is a stage in the plant's life cycle that is subject to intense competition and selection.

8.5 Vegetative growth

Although considerable differentiation has already taken place within the seed as germination begins, clearly as the root and shoot begin to grow the young plant begins to take on its characteristic form, even at quite an early stage. The forms that each plant takes must depend partly on internal genetic factors and partly on environmental factors, some of which will be internal and some external in origin. We know that even as early as the fertilized egg stage the polarity of the zygote may be determined by environmental conditions such as pH, temperature and oxygen concentration. However, as the emergence of root and shoot begin, the interplay of these factors with the internal, cellular, physical and chemical factors must be the way in which differentiation is effected.

Differentiation in the root

The root of the angiosperm has been extensively studied in the context of development. As a general rule it is safe to assume that the genetic constitution of most of the cells of the root is the same. This can be deduced from tissue culture experiments in which, provided the proper nutrients are supplied (see p. 168), many tissues and cells are capable of producing new plants. The problem is to find out why these genetically apparently identical cells are able to become so markedly different. As growth proceeds in the root apex, a number of definite phases of differentiation are clearly visible. Near the apex there is the apical meristem which is protected by the root cap (see fig. 8.6); this is the main site of cell division, though at the centre of the area there is a metabolically less active zone called the *quiescent centre*. Here division is less frequent but the cells may be

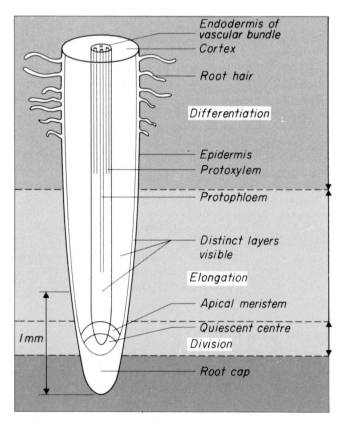

Fig. 8.6. Diagram of the root tip of the flowering plant showing the main areas of differentiation.

regarded as a reserve or pool for the replacement of the more rapidly dividing cells of the apex. It may also be a hormone source.

Behind the apex the principle development concerns cell elongation, though some differentiation is probably beginning and it is possible to distinguish a number of layers that are eventually to give rise to the fully differentiated structures of the mature primary root. The outer of these layers gives rise to the epidermis and the root hairs, the next forms the cortex and the inner develops into the vascular bundle.

Some 3–5 mm or so behind the apex differentiation is becoming clear; not only are the root hairs found but also the vascular bundle is beginning to be recognizable so that by the time the root hair has withered, perhaps 20 mm behind the apex, the vascular system takes on its mature primary structure.

Efforts have been made to try to gain an understanding of the reasons for this remarkably constant scheme of differentiation. Simple staining with methyl green pyronin (see appendix p. 242) shows that the DNA is most concentrated at the apex and also shows that the RNA, which stains green, is much more abundant above the lower 2 mm or so of the apex. During the phase of elongation that coincides with this increase in RNA, protein synthesis is taking place at an increasing rate, presumably, as it were, following the instructions of the RNA (see fig. 8.7). As would be expected, this is paralleled by an increase in the concentration and activity of certain enzymes, the most important of which are probably those concerned in the laying down and modification of the cell wall constituents. For instance, the enzyme invertase which is responsible for the conversion of sucrose into monosaccharides (see p. 137) is formed in large quantities during this phase.

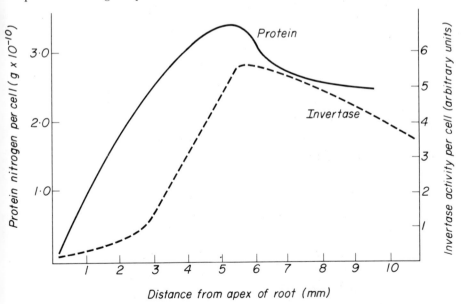

Fig. 8.7. Changes in protein and enzyme content during cell expansion. (Modified after Street, H. E. and Opik, H. (1970). *The Physiology of Flowering Plants*, London, Edward Arnold, and Black, M. and Edelman, J. (1970). *Plant Growth*, London, Heinemann.)

This sequence involving DNA, protein and enzyme still raises the question as to why roots produce certain enzymes in the first place; what determines that the root is a root with these various differentiating structures? The answer to this may well lie in the mechanisms of hormone action. If we assume that the root cell contains the full complement of genes, then some genes, such as those which would be needed for the formation of leaves or flowers, must be 'switched off' whilst others must be enabled to act. Following the lead of Jacob and Monod's work with bacteria we are beginning to gain some idea as to how such substances may act as *inducers* or *repressors* of gene action.

In the discussion of plant hormones in chapter 7 (p. 154) it was shown that the hormones which are likely to be involved in differentiation in the root apex are the auxins, cytokinins, gibberellins and ethene.

Taking the auxins for a start, the situation in the root is complex. Much of the IAA arriving at the root may be derived from the stem apex; indeed IAA exhibits strong polarity in its movement. However, the root apex itself may make other similar hormones, there being increasing evidence that these are not IAA, though chemically related to it.

There are three possible ways in which the auxins may affect the root cells; first they may be involved in switching-on some part of the RNA synthesis system so that a particular sequence of development follows. Second, the auxins may have a more direct effect on the plasticity of cell walls in the elongating zone (see p. 162) and finally we know that IAA has a considerable effect on the formation of roots from callus tissue (see p. 168) though this may be the result of the first possibility. Evidence regarding the possible primary role of auxins in determining RNA synthesis has come from a series of investigations involving the use of inhibitors which inhibit various stages in the RNA-enzyme-elongation of cell sequence. IAA undoubtedly has a role in determining more complex patterns of differentiation; mature primary roots have a characteristic pattern of xylem groups and it is known that IAA can have a modifying effect on the number of groups although these are normally standard for a particular species. Interesting work on the role of IAA in influencing the development of vascular

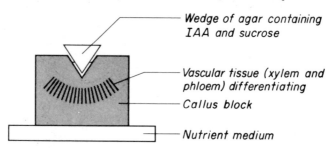

Fig. 8.8. IAA and the induction of vascular tissue in callus.

tissue has been carried out with callus tissue. Application of IAA together with sucrose causes the production of xylem and phloem within the callus chunk (see fig. 8.8).

These various roles of IAA and related auxins indicate that this is a hormone of exceeding importance in the development of the root. Less is known about the role of the other three hormone groups, the cytokinins, gibberellins and ethene. The cytokinins are synthesized near the apex and are known to have stimulating effects on both cell division and protein synthesis. There may also be interactions with other hormones such as IAA; as has been mentioned on page 168, root formation is favoured by relatively high concentrations of IAA and low concentrations of cytokinin. The role of gibberellic acid in differentiation may be similar to IAA in that it may also be responsible for the switching-on of a particular gene. It is also known to have an effect on membrane permeability and in this way may alter the rate at which both hormones and chemical substances in general are able to move through the tissue. It may be that it is through this that it is able to cause rapid elongation of cells through making more IAA or auxin available as well as having cell elongating properties on its own behalf. Finally ethene may be important in influencing the radial growth of the root.

Differentiation in the stem apex

The structure of the stem apex differs from that of the root in that there are no clear layers of tissue at the apex. There is a general area of dividing cells, the outer part of which, called the *tunica*, is composed of a number of distinct layers; these give rise to the leaf primordia. The inner, apparently more randomly arranged, cells inside the tunica are called the *corpus* and give rise to the parenchymatous cells of the central pith (see fig. 8.9).

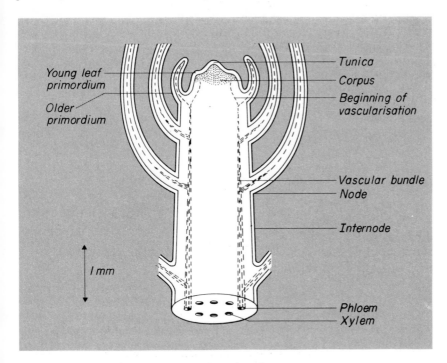

Fig. 8.9. Diagram of the stem apex of a flowering plant.

The leaf arrangement or *phyllotaxy* of a particular species is characteristic and flowering plants exhibit many different forms, various spiral systems being the most common. These arrangements depend on the initiation of primordia at the extreme apex and a good deal of experimentation and speculation has gone on in efforts to try to find out what determines a particular pattern. Exacting

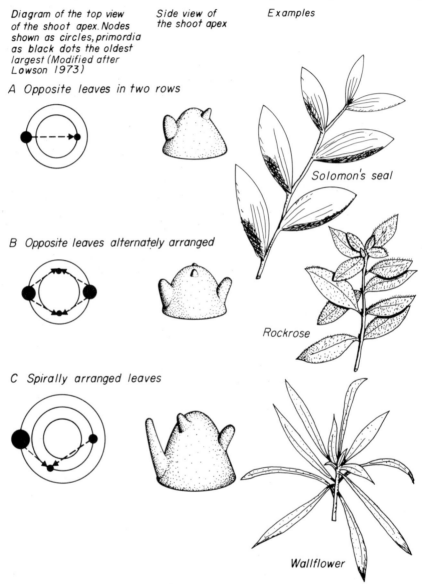

Diagram of the top view of the shoot apex. Nodes shown as circles, primordia as black dots the oldest largest (Modified after Lowson 1973)

Side view of the shoot apex

Examples

A *Opposite leaves in two rows*

Solomon's seal

B *Opposite leaves alternately arranged*

Rockrose

C *Spirally arranged leaves*

Wallflower

Fig. 8.10. Initiation of leaf primordia. Existing primordia inhibit the formation of new primordia, which are therefore formed as far distant as possible. Examples: A, Solomon's seal (*Polygonatum*): B, Rockrose (*Helianthemum*): C, Wallflower (*Cheiranthus*). (Lawson, 1973.)

operations in which primordia were removed showed that the next primordium to be formed usually appeared furthest away from the previous ones; it is as though each primordium tends to inhibit the production of others in its own vicinity (see fig. 8.10). This seemed to suggest that the determination of one primordium depended on substances diffusing from existing primorda. We have little idea as to what the morphogenic substances are, though the hormones discussed in connection with root differentiation are probably implicated.

If there is scant evidence as to what initiates the formation of leaf primordia it is hardly surprising that we do not know what causes the shape of leaves, though here the external environment has a much more profound effect. Most plants have cotyledons which are distinctly different from adult leaves, but leaves on young plants often differ from those on older plants and light, water, mineral availability and disease can all affect the shape and dimensions of the leaf. This is particularly obvious in some aquatic plants which may have finely dissected submerged leaves but relatively simple floating forms on the water surface.

On the other hand we do know that plants growing under high light conditions produce stems with relatively short internodes. This is probably due to a reduction of auxin brought about by the absorption of light. Whether the plant is of a simple monopodial, single stem growth form or branching and bushy depends on the degree of control or *apical dominance* exercised by the main apex. In the former the apex produces auxins which inhibit the growth of lateral buds while in the latter the concentration at the lateral buds is insufficient to cause inhibition and they also grow. The decapitation of the apex of a monopodial plant usually results in the lateral buds all beginning growth until one 'wins' and reasserts apical dominance. Apical dominance is clearly seen

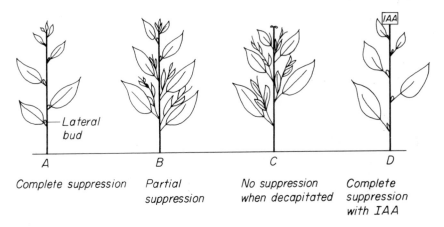

Complete suppression — Partial suppression — No suppression when decapitated — Complete suppression with IAA

A — B — C — D

Fig. 8.11. Apical dominance. (Redrawn after Black, M. and Edelman, J. (1970), *Plant Growth*, London, Heinemann.)

in many conifers, where, if the main apex is destroyed, perhaps in a storm, one of the lateral shoots takes over. Plants vary in the degree to which they exhibit apical dominance; some show only partial dominance, but in most cases apical dominance is restored if a block of agar impregnated with IAA is placed on the decapitated apex (see fig. 8.11).

Apical dominance is probably a much more complex phenomenon than these simple experiments with IAA would suggest. In the first place the concentration of IAA required to substitute for the apex is higher than would normally occur in the intact plant and secondly because most of the other hormones are probably also involved. Erect and prostrate genetic races of various plants, particularly conifers, are well-known in horticulture and illustrate again how relatively small changes in gene constitution can have profound effects on the growth form of the plant through influencing the hormone balance.

The onset of vegetative reproduction in many plants is another aspect of their life that depends to a large extent on hormones. Plants which have a spreading habit will be more likely to form creeping stems that root adventitiously; such stems must show diageotropism, that is they must orientate themselves at right angles to the force of gravity. The production of roots by such creeping stems when they come near to the moister layers of the soil also suggests that hormones such as IAA are stimulating the residual meristematic layers of the stem, the cambium and the pericycle, to differentiate. As so many vegetative reproductive systems are also linked with perennating systems, adaptations towards these ends rate highly in plant competition and survival.

8.6   The onset of sexual reproduction

Under natural conditions plants vary greatly in the age at which they reach sexual maturity. In the flowering plants some annuals such as the shepherd's purse (*Capsella bursa-pastoris*) and the groundsel (*Senecio vulgaris*) may be able to reach flowering in 14 days after the seed is shed and die as soon as their seed is all ripened. Other *monocarpic* plants include biennials such as some strains of

Fig. 8.12. *Saxifraga longifolia* in the Pyrenees. An example of an alpine monocarpic plant. (Photograph by W. Schacht, Courtesy Alpine Garden Society.)

henbane, beetroot and the Canterbury bell (*Campanula medium*) which make considerable vegetative growth during their first year and then flower in the second, usually to produce a vast quantity of seeds, only to die soon after. Some other plants may still be monocarpic but take several years to flower. Good examples are the century plant (*Agave americana*) and the alpine *Saxifraga longiflora* (see fig. 8.12). Polycarpic perennials may take one or many years to reach flowering size though many gardeners know that some perennials may die if they are allowed to flower too soon before becoming properly established. In this respect the distinction between garden biennials and perennials such as the wallflower and stock are not well defined.

For most plants to begin sexual reproduction there must be sufficient food available to furnish the reproductive parts and swell the seeds (or spores). Annuals clearly do this at the expense of the rest of the plant and it is not unusual for some perennials to literally 'flower themselves to death' if the requirements of flowers and fruits are too great. Sometimes too, diseases can cause a premature maturing of the plant. It would appear then that provided the nutritional requirements of the plant are met, flowering can take place at the appropriate season.

8.7 The Phytochrome System

It has been known since the work of Garner and Allard in the 1920s that plants could be classified into three main groups according to the day-length conditions that they require for flowering. This phenomenon is known as photoperiodism (see fig. 8.13). Plants that require long days and short nights to induce flowering are called long-day plants; the reverse are short day plants, requiring a maximum of fourteen hours daylight; the third group are neutral or indifferent and may flower throughout the year. The table below lists a few plants in each group:

Day length	Examples
Short	*Chrysanthemum*, cocklebur (*Xanthium*)
	Salvia, Maryland Mammoth tobacco
Long	*Petunia*, barley (*Hordeum*), spinach
Neutral	tomato, dandelion (*Taraxacum*)

This division of plants follows closely their geographic distribution. Long-day plants, which are normally summer flowering, are frequently found in Arctic and northern areas, where the summer is short and the day length very long indeed, while the short-day plants are more common nearer the Equator, and tend to be autumn or spring flowering.

The problem is, how does day (or night) length affect the plant so as to change the normal leaf primordia of the stem apex into flowering primordia? In short-day plants evidence that it is the dark period rather than the light period which is important comes from experiments in which plants, treated with the appropriate day length, are given a short bright light flash in the middle of the night; this inhibits flowering. This phenomenon has been used commercially to retard the flowering of autumn-flowering *Chrysanthemum* so that they bloom at Christmas.

The number of cycles of day plus night that are required to cause flowering is quite small; cocklebur (*Xanthium*) requires only one cycle, but most plants require about ten.

In long-day plants it can hardly be the dark period which is important, as they will flower in continuous daylight, so there must be different systems operating in the two groups.

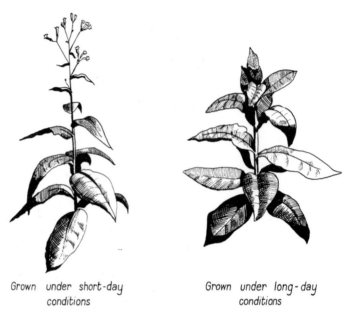

Grown   under   short-day   Grown   under   long-day
conditions   conditions

Fig. 8.13. The effect of day length on flowering of tobacco (Maryland Mammoth). (From *Principles of Plant Physiology* by James Bonner & Arthur W. Galston. San Francisco: W. H. Freeman & Co., 1952. *After* Garner & Allard: *Yearbook of Agriculture*, 1920.)

Considerable work has been carried out recently on the nature of this photoperiodic induction. In the first place it was found that only red light (660 nm) was effective in inhibiting the flowering of short-day plants when they received a light flash in the middle of the night. The next step came when it was found that any inhibition could be reversed if the plants were treated with far-red light, at 730 nm. Indeed in a manner similar to the effect of light on germination (see p. 180) the various treatments could be alternated many times and flowering would still result if the final treatment was with far-red light:

Red light (660 nm)   Vegetative
Far red light (730 nm)   Flowering
Red, then far-red, then red   Vegetative
Red, then far-red, then red, then far-red   Flowering

As in germination studies this reversibility suggests that the single photoreceptive compound, *phytochrome*, is involved. The pigment can exist in two forms; P_r and P_{fr} (i.e. Phytochrome red and Phytochrome far-red). Normal daylight contains more orange-red light of wavelength 660 nm; the P_r form of the pigment absorbs light at this wavelength and the pigment changes to the P_{fr} form. This latter form can be converted back into the P_r form by far-red light, a process which is more likely to occur at dusk and dawn, or by slow temperature-sensitive chemical changes. The diagram below summarizes these relationships:

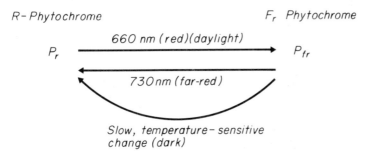

In short day plants then, we have a system in which flowering will take place eventually if sufficient R-Phytochrome (P_r) accumulates. The question is, how does the formation of this pigment, which is known to be formed in the active green leaves and is itself unlikely to be translocated, result in the transmission of some substance to the stem apex to effect the conversion of normal leaf primordia into flowering parts? That some hormone is involved is shown from simple experiments with short day cocklebur (*Xanthium*) which will produce flowers even if the only part of the plant receiving short day treatment is a single leaf some distance away from the developing flowers. Grafting experiments have confirmed this (see fig. 8.14); if a plant that has been previously treated so as to initiate flowering is grafted on to a plant growing under inappropriate day length conditions, then flowering will result. This hypothetical flowering hormone although not so far isolated has been called *florigen*. It is also interesting

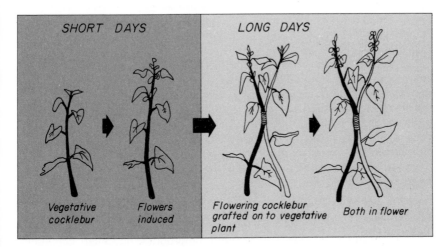

Fig. 8.14. Transmission of the flowering stimulus across a graft union. (Adapted from Hamner, K. C. and Bonner, J. (1938). *Bot. Gaz* 100, 388–431.)

that many different plants are capable of inducing flowering in completely different species; for instance a treated leaf of tobacco (*Nicotiana*) can be grafted on to henbane (*Hyoscyamus*) and cause flowering. Although this seems to suggest that a single substance is involved the situation is not at all clear cut.

In efforts to identify the hormone or hormones involved particular attention has been given to the gibberellins. They are found particularly in long day plants and may initiate flowering in that group. This has been demonstrated quite clearly with brookweed (*Samolus*); when this long day plant is grown under short day conditions it can be induced to form flowers after the application of GA. Indeed up to a point, the more GA that is applied the greater the number of flowers formed (see fig. 8.15).

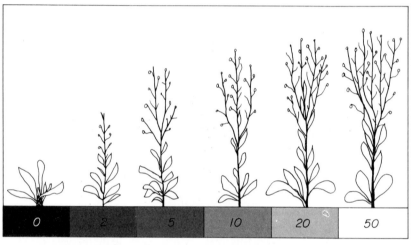

Concentration of GA applied (μg per plant per day)

Fig. 8.15. Induction of flowering by gibberellic acid when applied to the long day plant, *Samolus*, grown under short days (9 hours). (Redrawn after Lang, (1957) *Proc. Nat. Acad. Sci.*, 43, 709–717.)

GA does not seem to be involved in the flowering of short day plants but it is known that other hormones such as abscisic acid (ABA) are formed under short day conditions and this substance may be implicated. The effect of red light on the phytochrome system may be to alter the permeability of membranes and it could be that it is through this that the appropriate balance of the hormones needed to induce flowering is achieved. However, the fact remains that often only a remarkably small number of photoperiods of the correct kind are sufficient to induce plants to flower and so ensure that their reproduction is synchronized properly with the seasons.

8.8   Pollination and fertilization

Given the required level of maturity and the appropriate day length conditions, the leaf primordia become modified to flower primordia and some weeks later the flower will open. Anthocyanidins may have accumulated in the petals and sugars are often produced in nectaries, while pollinating insects may also be attracted to the flower by the high protein and nutrient level of the pollen itself.

The open petals of many flowers may also serve to focus radiant heat; a factor which may be very important in cold climates where the activity of insects would otherwise be rather low.

Meiosis takes place in the formation of the pollen and, through the agency of some pollinating agent, pollen grains arrive on the stigma of the female parts of of the flower (see fig. 8.16). Leaving aside the details of these important

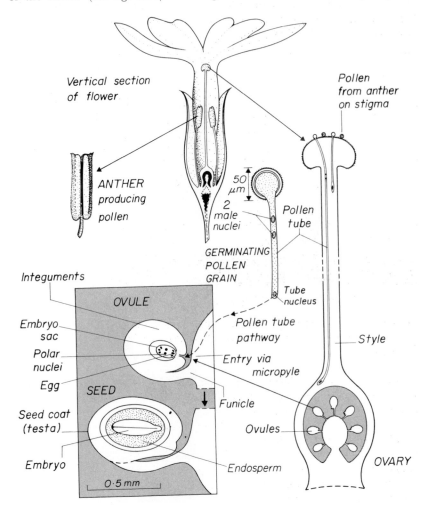

Fig. 8.16. Sexual reproduction in *Primula*. A long styled (pin-eyed) flower is shown, together with anther and ovary, ovule and seed.

steps there are a number of physiological aspects with regard to fertilization that are worth examining more fully. The first concerns the germination of pollen. Although some pollen grains of the wrong species undoubtedly germinate on the stigma, it may be supposed that both structural and chemical interactions have to take place for pollination to be successfully followed by fertilization. Most pollen grains have a particularly characteristic shape, often with various echinations on the outer wall, though related species are often very alike. Similarly the stigma may have various spines as in that of *Primula*, and *Crocus* (see fig. 8.17). Scanning electron micrographs have shown how elaborate and

50 μm

Fig. 8.17. Pollen grains sticking to the microscopic papillae on the surface of the stigma of *Crocus*.

attractive these various sculpturings can be and it is tempting to suggest that the first interaction between pollen and stigma is one of mechanical fit. However we do know from investigations of the conditions required for the germination of pollen (see appendix p. 251) that it is quite easy to germinate pollen in artificial conditions provided sucrose and boric acid solution are provided. Their germination and pollen tube formation can be watched very easily using the hanging-drop technique (see fig. 8.18).

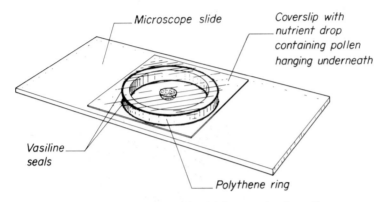

Fig. 8.18. The hanging drop technique for examining germinating pollen.

As the pollen tubes force their way down the stigma, probably assisted by the production of their own enzymes, nutritional and probably other kinds of interaction may take place. Even if the pollen tube is of the correct species and the compatibility strain happens to be wrong, then pollen tube growth is slower and less likely to be effective than if the 'correct' tubes are growing down the stigma. This occurs in *Primula* in the event of self-fertilization. In members of this genus that have species with differing style length (heterostyly), pollen tube growth is best where the pollen is derived from flowers having differing style length from the recipient. Fig. 8.16 shows a long-styled or *pin-eyed* form; pollen received from a short-styled or *thrum-eyed* flower will be most successful in this case. Investigation of the growth requirements of pollen grains is important as it has led to the discovery that, provided appropriate nutrients and hormones such as the auxins and cytokinins are given, complete haploid plants can be produced (see p. 169).

As the pollen tube grows down the stylar tissues and approaches the ovary it is thought that chemical substances, produced by the ovules begin to exert an attracting effect on the pollen tube so that its growth is directed towards the ovule stalk (funicle) and finally into the embryo sac, often, but not always through the micropyle. On entering the embryo sac one of the male nuclei or sperms fuses with the egg and the other with the two polar nuclei to form a triploid fusion nucleus.

During the period that the flower is capable of fertilization there may be a period of some days when the petals open and close in a set diurnal rhythm. Closure is more usual at night and serves to protect the flower when pollinating insects are absent, though both light and temperature may be involved. Some flowers e.g. *Crocus* are highly sensitive to temperature changes and will open their flowers in a few minutes if brought into a warm room. When the flower is young the movements are clearer; the petals being tightly closed over the flower parts or alternatively widely opened. After pollination and as the flower parts begin to age, the movements become less obvious and the petals may remain fully open or closed until they wither or fall off. These *nastic movements* may be due to changes in turgidity of the cells of the petals or, as is the case in *Crocus*, due to differential growth rates between the inside and outside of the flower petals.

Apart from the petals, other parts of the flower such as the stamens and stigma may also show movements, but these are often related to fertilization, bringing either anther or stigma into the line of approach of the pollinating insect.

8.9   Fruit and seed formation

The fertilized egg or zygote then develops into the embryo, while the triploid fusion nucleus divides to form the endosperm, the food reserve of the seed. This is not always present on the ripe seed, as the food substances it has contained are often taken up by the developing embryo (see fig. 8.19).

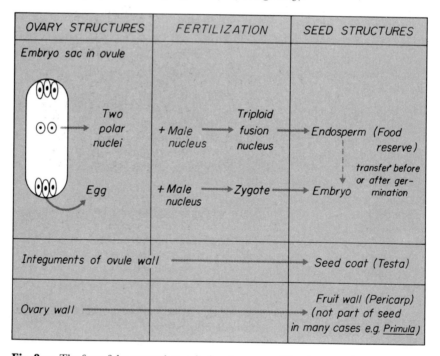

Fig. 8.19. The fate of the ovary tissues during the stages of development leading to seed formation.

The integuments may form the seed coat (testa) or the ovary wall may continue to surround the individual ovule so that there are more layers around it. Quite commonly this ovary wall or *pericarp* becomes fleshy and one or a number of seeds may be shed within the single *fruit*, as in the tomato. Fruit growers have known for long enough of the importance of pollination for fruit formation and it is known that materials diffusing out of the pollen as well as from the developing seed itself both produce growth substances which stimulate the development of the fruit (see fig. 8.20). Auxins, gibberellins and cytokinins are probably all

Fig. 8.20. Effects of developing achenes on the growth of the strawberry receptacle. (1) Unpollinated flower: no development of receptacle. (2) One pollinated achene: growth of the receptacle around it. (3) Several pollinated achenes: several areas of receptacle growth. (4) Many pollinated achenes. (From Nitsch, 1965. *Encyclopedia of Plant Physiology*, ed. Ruhland 15 (1), 1601. Springer-Verlag, Berlin.)

involved here and knowledge of this has led to techniques to stimulate fruit formation in unfertilized flowers so that pipless fruits are formed. On the other hand, there are of course numerous cultivated forms of fruits such as orange, tangerine, grapefuit and grape which may still be able to form fruits without any seeds at all.

Considerable metabolic activity takes place after fertilization to allow for the production of the various substances in the seed and fruit, but as ripeness is achieved, metabolic activity, at least of the embryo and endosperm, declines to virtually nothing. At this time characteristic changes may be taking place if the fruit is a fleshy one, plant acids such as 2-hydroxybutanedioic acid which have made the fruit bitter and inedible are gradually removed and sugars accumulate. Further changes may take place usually, but not entirely, in the outer parts of the fruit where red anthocyanidin and yellowish-orange carotenoid pigments (e.g. lycopene in tomato) may accumulate. These are likely to be obvious to birds and mammals and the fruits will provide them with a useful supply of food in the autumn as well as helping to disperse the seeds.

8.10   Senescence and dormancy of buds

Different groups of plants reach maturity at different times and after flowering and fruit formation may either die as in monocarpic plants or continue alive for many years flowering year after year (see p. 189). However even perennial plants usually show some annual indications of ageing or *senescence* in the gradual

Fig. 8.21. The Grizzly Giant in the Mariposa grove of Big Trees (*Sequoia gigantea*). Age: 3800 years, height: 64 m, basal diameter: 10.7 m. Although a number of branches appear to be senescent or have been damaged by wind or lightning, others appear to be healthy and to have retained their vigour. (Courtesy of Mrs. C. R. Baron.)

slowing of their growth and their annual fall of leaves. These perennials often have a given life-span, albeit a variable one; they may become senescent through some nutritional factor in the soil being limiting to their growth or they may become increasingly susceptible to damage from other environmental factors such as wind and disease though in some instances such as the aged bristle cone pine (*Pinus aristata*) and the Big Tree (*Sequoia gigantea*) this may take thousands of years (see fig. 8.21). Clearly, in monocarpic plants the acts of flowering and fruiting impose heavy strains on the plant's resources so that there is a great demand for the products of photosynthesis. A similar point is well-known to gardeners who 'pinch-out' the flowers of a perennial plant if it starts to flower before the plant is well-established and has sufficient resources to provide both for reproduction and continued growth. But annuals and biennials, like some kinds of cells such as the xylem, must in a sense be *programmed to die* and some ageing factors, directly or indirectly under genetic control, must be involved. These could result from a decline in the rate of formation of RNA and synthesis

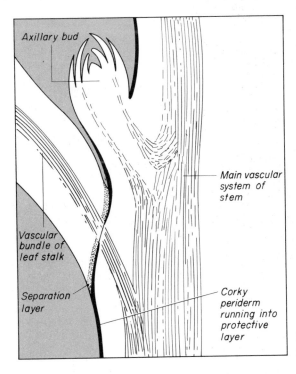

Fig. 8.22. Abscission zone in the Maple (*Acer*) leaf stalk.

of protein so that tissue growth and replacement cannot be maintained and plant vigour is gradually lost.

One of the senescence phenomena that has been quite fully investigated is that of leaf-fall or *abscission*. As autumn approaches the changes in leaf colour that have already been described (see p. 146) may occur and anatomical changes within the leaf stalk (petiole) are also taking place (see fig. 8.22). In the abscission zone, the cells of the proximal zone, nearest the stem take on a protective function and become corky or suberized. Distally the cells become progressively weaker in the *separation layer* as the calcium pectate and cellulose materials of their walls are broken down due to the activity of pectinase and cellulase enzymes. As the time for leaf fall approaches the xylem vessels become plugged with intrusions called *tyloses*, from their walls. Finally wind and frost often combine in the dislodging the leaf. Rather similar changes take place in the fall of flower petals and fruits.

If a green leaf is removed from the parent plant and kept with its stalk in water it becomes yellow and senescent in a few days, but if the stalk is kept dipped in water containing 5 parts per million cytokinin (kinetin, 6-furfurylaminopurine), then the green colour of the leaves is maintained much longer. It is thought that the cytokinin helps to keep up the rates of RNA and protein synthesis so that leaf vigour is not lost. Evidence of the involvement of other growth hormones such as the auxins (IAA) has come from experiments in which the leaf blade is removed and the stump treated with the hormone. When this is done without hormone treatment abscission of the remaining leaf stalk follows within a few days (see fig. 8.23). Application of IAA in lanolin to the stump (see appendix p. 252) often prevents the formation of this abscission zone.

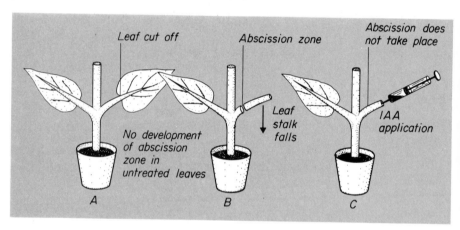

Fig. 8.23. IAA and abscission. Removal of the leaf blade results in abscission of the leaf stalk some days later. However, application of IAA to the stump after the leaf is cut off suppresses the formation of the abscission zone.

It looks then as though both the cytokinins and auxins may help prevent the onset of senescence and maintain the leaf in its vigorous metabolic state. On the other hand abscisic acid (ABA) if applied to stumps in a similar way usually promotes abscission, just as the gas ethene may promote yellowing in both leaves and fruits. It is tempting to suppose that these latter two hormones act as antagonists to cytokinins and auxins though no doubt the situation is much more involved. Ageing of a particular plant or part of a plant must depend on a complex of genetic and environmental factors which allow for the accumulation of a particular hormonal balance at the appropriate time of year.

Rather similar changes in the balance of hormones are probably involved in the formation of *dormant* structures such as winter buds. Once again it must be the environmental conditions of winter and spring which trigger hormonal changes so that the rate of growth is diminished and then increased. This capacity of the plant to synchronize its life to the seasons and to some extent anticipate the forthcoming changes in the season is one of the more important of plant adaptations.

Further reading on Physiological Organization and the Life Cycle

BLACK, M. and EDELMAN, J. (1970). *Plant Growth.* London, Heinemann.

GALSTON, A. W. and DAVIES, P. J. (1970). *Control Mechanism in Plant Development.* Engelwood Cliffs, Prentice-Hall.

HILL, T. A. (1973). *Endogenous Plant Growth Substances.* Institute of Biology's Studies in Biology no. 40. London, Edward Arnold.

JOHN, B. and LEWIS, K. R. (1972). *Somatic Cell Division.* Oxford Biology Reader No. 26, Oxford University Press.

KEMP, R. F. O. (1970). *Cell Division and Heredity.* Institute of Biology. Studies in Biology no. 21. London, Edward Arnold.

KENDRICK, R. E. and FRANKLAND, B. *Phytochrome and Plant Growth.* (1976). Institute of Biology's Studies in Biology no. 68. London, Edward Arnold.

MCLEISH, J. and SNOAD, B. (1958). *Looking at Chromosomes.* London, MacMillan.

VILLIERS, T. A. (1975). *Dormancy and the Survival of Plants.* Institute of Biology's Studies in Biology no. 57. London, Edward Arnold.

WARDLAW, C. W. (1968). *Morphogenesis in plants.* London, Methuen.

Appendix A. Experimental Procedures

Many of the techniques and materials used in these sections require careful preparation and handling; make sure you are aware of the various hazards and use appropriate safety procedures that may be laid down by the Local Education Department and the Health and Safety at Work Act.

An asterisk (*) indicates that the reagent is described in Appendix B.

APPENDIX TO CHAPTER 2

1. Determination of the osmotic potential of the cell sap at incipient plasmolysis
2. Determination of the water potential of the cell by the strip method
3. Determination of the water potential of the cells of bulky tissues by the weighing method
4. Determination of the effect of temperature on the permeability of the cytoplasm
5. Determination of the transpiration rate together with measurements of the environmental conditions and evaporation rate
6. The nail-varnish replica technique for examination of the leaf or stem epidermis
7. Measurement of the resistance to air flow offered by the leaf using Meidner's porometer
8. Investigation of the accumulation of potassium in the stomatal guard cells

1. Determination of the osmotic potential of the cell sap at incipient plasmolysis (see p. 16)

Materials. Staminal hairs of *Tradescantia* and *Zebrina*; epidermal strips of leaves, especially those coloured by anthocyanins (*Rhoeo discolor, Tradescantia, Zebrina*); epidermis of red onion bulb; leaves of *Elodea* and mosses with only one cell layer.
Method. Prepare a series of solutions of sucrose of different molarities: 0.5, 0.4, 0.3, 0.2, 0.1 M or higher molarities, and distilled water. Place a small piece of the tissue being examined in each of these solutions in small, covered cavity dishes. Leave for twenty minutes and then mount the tissue on a slide in the same solution and cover with a cover-slip. The preparation in which about half the cells are just starting plasmolysis is regarded as at *incipient plasmolysis* and the internal and external solutions are considered isotonic.

A molar solution exerts an osmotic pressure of 2269 kPa (2.27 MPa) or 22.4 atmospheres at N.T.P. provided that no ionization of the plasmolysing solution takes place. Sucrose solution, unlike potassium nitrate, is un-ionized, so no correction for ionization of the solution is needed in calculating the osmotic

Molarity	O.P./Atmosphere	O.P./kPa	Molarity	O.P./Atmosphere	O.P./kPa
0.05	1.3	130	0.55	16.0	1620
0.10	2.6	260	0.60	17.8	1800
0.15	4.0	400	0.65	19.6	1900
0.20	5.3	540	0.70	21.5	2180
0.25	6.7	680	0.75	23.4	2370
0.30	8.1	820	0.80	25.5	2580
0.35	9.6	970	0.85	27.6	2800
0.40	11.1	1120	0.90	29.7	3000
0.45	12.7	1290	0.95	32.1	3250
0.50	14.3	1450	1.00	34.6	3500

Osmotic Pressures of different molarities of sucrose solutions at 20°C.
Data modified from Ursprung and Blum. *Ber. Deutsch. Bot. Ges.*, **34**, 525–554, 1916.
(From Meyer and Anderson's *Laboratory Plant Physiology.* © 1955, D. Van Nostrand Co.,
Inc., Princeton, N.J.)

potential of the cell sap. The table above gives the osmotic pressure in
atmospheres for different molarities of sugar solution at 20°C.

2. Determination of the water potential of the cell by the strip method
(see p. 19)

Material. Thin strips of beetroot or of the corona of the daffodil.
Method. Cut the strips exactly 30 mm long and as far as possible identical. Place
these in a similar range of solutions to those described above in exp. 1, in covered
Petri dishes or test-tubes. Leave for at least an hour and measure again as
accurately as possible. Plot on a graph the ratio of the initial to final length,
against the molarity. Estimate the water potential of the tissue, knowing that a
ratio of one indicates the molarity of the cell sap. Determine the water potential
in atmospheres by referring to the above table.

3. Determination of the water potential of the cells of bulky tissues by
the weighing method (see p. 19)

Material. Beetroot or potato.
Method. Use a cork borer to cut discs of the material so as to produce six
equivalent portions, each of about 2 g. Dry and weigh each accurately and place
separately in the range of solutions as used in exp. 1. Cover up the containers and
leave for twenty-four hours. Dry quickly with blotting paper, reweigh and
calculate the water potential as in exp. 2.

4. Determination of the effect of temperature on the permeability of
the cytoplasm (see p. 20)

Material. Beetroot.
Method. Cut six slices of beetroot, 30 mm long, 5 mm wide and 2 mm thick,
similar as far as possible. Wash these in running water overnight and transfer

them to a beaker of distilled water. Heat 200 cm^3 of distilled water to 75°C. and immerse one of the slices in the heated water for exactly one minute, then transfer it to a test-tube containing 10 cm^3 of distilled water at room temperature. As the beaker of distilled water cools repeat the procedure with a fresh slice of beetroot at 70°, 67°, 65°, 63°, 60°, 55° and 50°C. In each case place the treated slice in a separate tube of distilled water.

After half an hour shake the tubes and compare the relative amounts of red pigment (anthocyanidin) that have diffused out of the slices in each case. If possible compare the colours against a colour chart or measure their density using a photoelectric colorimeter. Plot the colour density against the temperature on a graph and discuss how temperature affects the permeability of the cytoplasmic membrane. What is the lowest temperature that destroys the physical nature of the cytoplasm and permits the outward diffusion of anthocyanidin?

(From Meyer and Anderson's *Laboratory Plant Physiology.* © 1955, D. Van Nostrand Co., Inc., Princeton, N.J.)

5. Determination of the transpiration rate together with measurements of the environmental conditions and evaporation rate

Materials. Mesophytes e.g. dog's mercury (*Mercurialis* spp.) or hairy willow herb (*Epilobium hirsutum*) and succulents such as *Bryophyllum daigremontianum*.

Method. One of the best methods for determining the transpiration rate involves the continuous accurate weighing of a single leaf, using sensitive balance (see p. 22). Evaporation rates can be similarly determined using moistened filter paper.

However, the potometer and atmometer (see p. 21) may be more generally available. These can be modified for continuous operation by connecting a vertical glass tube to the end of the capillary measuring tube by means of a piece of PVC tubing. The end of this tube is dipped into a beaker of water so that the plant may transpire continuously (or the porous pot evaporate) without air being drawn into the system. A small air bubble is held in the PVC tube; this can be introduced into the capillary by squeezing the tube. Once the measurement of the rate has been taken, the bubble can be driven back into its place in the connecting tube by opening the tap on the reservoir. The usual care must be taken in setting up the apparatus to exclude leaks and air locks.

Simultaneous measurements of the temperature, light intensity and, if necessary, wind velocity can also be taken. Ideally the rates of evaporation and transpiration per unit surface area should be calculated. The evaporating surface of the porous pot of the atmometer is easily calculated from the expression $\pi r^2 + 2\pi rh$, but to obtain the transpiring surface of the whole plant it is necessary to remove all the leaves and, by drawing their outlines on a piece of graph paper, find their total surface area. This is not quite so laborious as it sounds, as the leaves can usually be sorted into a small number of size groupings which considerably speeds up the calculation.

The various rates may be compared on a graph alongside plots for the temperature and light intensity, and conclusions can then be drawn regarding the effect of environmental conditions on the rate of transpiration.

6. The nail-varnish replica technique for examination of the leaf or stem epidermis (see p. 26)

Materials. Plants from which it is difficult or undesirable to remove the epidermis, particularly the stems of cacti and succulent plants.
Method. Paint a little clear nail-varnish on to the leaf or stem. After a few minutes remove the varnish with tweezers and mount it in water on a slide and examine under the microscope. Count the number of stomata in a given area (e.g. the low-power microscope field of view).

7. Measurement of the resistance to air flow offered by the leaf using Meidner's porometer (see p. 25)

Materials. Leaves with stomata on both surfaces work best; e.g. runner bean and sunflower.
Method. A simple porometer is illustrated in figs. A.1 and A.2. This is, in effect, a Perspex clamp which fits over the surface of the leaf. Air can be sucked through the leaf by squeezing and releasing the rubber bulb of the pipette.

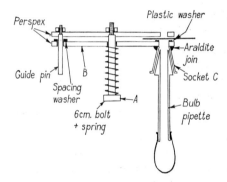

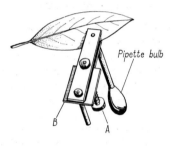

Fig. A.1. Side-view of Meidner's porometer.

Fig. A.2. Meidner's porometer attached to leaf.

The clamp can be opened to fix it to the leaf by placing the thumb on the bolt-head *A*, two fingers on the extended parts *B* of the clamp, and squeezing.

First remove the pipette, then open the clamp by placing the thumb on the head of the bolt *A* and two fingers on the extended parts of the Perspex plate *B*, and then pressing. Select a suitable piece of leaf which has no large veins and release the pressure, but continue to hold the clamp.

Squeeze the rubber bulb of the pipette and then insert it into the socket *C*. Release the bulb and take the time for it to inflate. This period is proportional to the resistance of the leaf and gives a measure of the degree of opening of the stomata. Consistent results are obtained quite easily with a little practice. The airtightness of the apparatus can be checked using a microscope coverslip in lieu of the leaf.

(Adapted from Meidner (1965). *School Science Review*, No. 161, p. 149.)

8. Investigation of the accumulation of potassium in the stomatal guard cells (see p. 27)

Materials. Leaves of broad bean (*Vicia faba*), *Commelina* and many other plants. One batch of leaves should have been kept in the dark, the other in bright light with minimum carbon dioxide.

Method. Treat each batch separately as follows: strip the epidermis from the lower surface of the leaf and immediately immerse it in an ice-cold solution of freshly made up sodium cobaltinitrite* for thirty minutes. Wash in ice-cold distilled water until all excess stain is removed, and then place the epidermis in a five per cent solution of yellow ammonium sulphide (*caution*: toxic vapour) for two minutes, rinse and examine. Compare the material from the light and dark, noting the orange precipitate which indicates the presence of potassium ions.

(After Mansfield, T. A. (1970). 'Stomata in new perspective'. *S.S.R.* No. 179, p. 316)

APPENDIX TO CHAPTER 3

1. Identification of reducing sugars (e.g. glucose, fructose and maltose)
2. Identification of non-reducing sugars (e.g. sucrose)
3. Identification of storage polysaccharides (e.g. starch and inulin)
4. Determination of the compensation period
5. Investigation of the factors affecting the rate of photosynthesis
6. Estimation of the percentage of carbon dioxide and oxygen in a gas sample (Eggleston's method)
7. General technique of chromatography
8. Filter-paper chromatography, using long strips
9. Filter-paper chromatography, using large sheets
10. Thin-layer chromatography
11. Extraction of the photosynthetic pigments
12. Separation of the chlorophyll pigments by column chromatography
13. Setting-up and calibration of the spectrometer
14. Investigation of the fluorescence of the chlorophyll pigments
15. Investigation of the catalytic properties of extracted chloroplasts (the Hill reaction)
16. Investigation of the uptake of carbon dioxide into the leaf, using ^{14}C radioactive tracer
17. Investigation of the synthesis of starch from glucose in the dark by leaves of *Pelargonium*

1. Identification of reducing sugars, e.g. the monosaccharides *glucose* and *fructose* and the disaccharide *maltose*

Materials. Most fruits, such as the apple and orange, onions, and green leaves (cleared of chlorophyll by filtration or centrifuging) (see p. 46)

Tests. These sugars reduce the Copper II ions of Benedict's* solution to Copper I oxide, give a red-orange or brown precipitate on boiling. Dilute sugar solutions give only slightly yellow or greenish colours, but by careful comparison with a non-sugar-containing control solution it is usually possible to detect even small quantities of sugar. By comparison with sugar solutions of different known strengths it is also possible to obtain a rough quantitative estimate of the amount of sugar present.

This test is probably best operated on bulky tissues and extracts, but, if necessary, can also be done on a micro-scale. Sections of the material should be boiled in the reagent for a few minutes in a watch glass; grains of cuprous oxide, appearing black by transmitted light under the microscope, will appear where reducing sugars are present. It is important, though, to compare these with a preparation that has been boiled in water.

2. Identification of non-reducing sugars, e.g. the disaccharide *sucrose*

Materials. As for reducing sugars.

Tests. This does not reduce Benedict's* solution direct, and it is first necessary to hydrolyse or 'invert' the sucrose by boiling it for about a minute with dilute hydrochloric acid. This breaks the disaccharide into its two monosaccharide units, glucose and fructose. After neutralizing excess hydrochloric acid by adding solid sodium carbonate until effervescence ceases, Benedict's test can be applied as for reducing sugars. No simple microchemical tests for sucrose are available.

3. Identification of storage polysaccharides, e.g. starch and inulin

These are composed of a large number of hexose units joined together to form chains of different types. Prolonged hydrolysis may result in the formation of reducing sugars, which may be detected as above. The simpler polysaccharides are essentially food reserve substances and the more complex are important structural constituents of the cell.

(a) STARCH

Materials. Potato tubers, many rhizomes, the endosperm of seeds and photosynthesizing leaves.

Test. Starch may be identified in bulk or on a micro-scale by the use of iodine solution (dissolved in potassium iodide*), which gives a blue-black colour. During the enzymatic hydrolysis of starch (see p. 139) simpler short-chain molecules called *dextrins* may be formed, the larger of which may react with iodine to give a violet colour.

(b) INULIN

Materials. The root tubers of *Dahlia*, the stem tubers of the artichoke (*Helianthus tuberosus*), the root of the dandelion.

Test. This is a storage polysaccharide similar to starch, but it does not react with iodine and is easily recognized under the microscope by the clusters of fan-shaped crystals which appear when the tissue is dehydrated with alcohol.

4. Determination of the compensation period (see p. 43)

Material. Any small plant, e.g. wall ferns and bryophytes; single leaves of many plants; shoots of aquatic plants (e.g. *Ranunculus, Elodea, Callitriche*).

Method. Place a filter-paper in the bottom of a small deep glass dish and moisten it with distilled water. Place a cavity block on the filter-paper and fill it with 1 cm³ sodium bicarbonate indicator solution.* This solution tends to equilibrate its carbon dioxide content with that of the atmosphere, and should previously have been aspirated through with atmospheric air until it is a red colour. Arrange the plant material, which should have a minimum of soil, around the cavity block and seal the dish with a vaselined piece of glass (see fig. A.3). Single leaves may be arranged as shown in fig. A.4, alternatively water plants may be placed direct in the indicator solution and the tube sealed.

Keep the dish or tube in a dark room at a fixed temperature for about ten hours so that respiration alone can take place, and then transfer it to bright illumination. Take care not to warm the plant with the illuminating lamps and keep it at the same temperature throughout the experiment. The compensation period is given by the time taken for the indicator to change back from yellow to the original red colour. The experiment may be repeated at different temperatures and at different light intensities.

Modified after Hosokawa & Odani (1957). Compensation period and vertical ranges of epiphtes. *Journal of Ecology*, **45**, No. 3, p. 901.

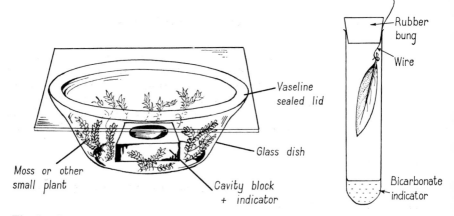

Fig. A.3. Jar system used for determining the compensation period of small plants.

Fig. A.4. Simple tube for determining the compensation period of a single leaf.

5. Investigation of factors affecting rate of photosynthesis (see p. 47)

Material. Young sprigs of *Elodea canadensis*.

Method. Place a few short sprigs of *Elodea* in a 1-dm³ beaker, preferably containing the water in which the plant was growing. Illuminate brightly for about half an hour. At the end of this period a stream of bubbles should be coming from the cut ends of the shoots. The number of bubbles produced per minute gives a measure of the rate of photosynthesis. The percentage oxygen in the gas evolved can be estimated if required (see exp. 6 below). Take care not to move the beaker or sprigs of *Elodea* during the course of the experiment and also give the material time to settle down when alterations are made to the conditions. It may be useful to use a piece of glass as a heat filter between the bulb and the beaker. The distance (d) between the bulb and the *Elodea* should be measured as accurately as possible from the front of the bulb to the centre of the photosynthetic area of the *Elodea*. The rate of bubbling will differ from shoot to shoot, so once a shoot has been selected it alone should be used throughout the experiment.

(a) To determine the effect of light intensity on the rate of photosynthesis

Vary the light intensity by moving a bright lamp progressively nearer the *Elodea*. Knowing that the light intensity at distance d from the lamp is proportional to $1/d^2$, plot $1/d^2$ against the amount of oxygen produced. Where the graph is a straight line, light is controlling the rate of the reaction, but if the rate of photosynthesis tends to show no further increase as the higher intensities are approached, then some other factor, possibly temperature or carbon dioxide availability, is limiting the rate of the reaction. The concentration of carbon dioxide can be increased by carefully adding a dilute solution of sodium bicarbonate from a pipette.

(b) To determine the effect of temperature on the rate of photosynthesis

Set up the apparatus as in the previous experiment, but keep the light intensity constant, first, at a low intensity and then at a high intensity. Record the rate of photosynthesis in each case at $18°$ and $28°C$. Under high illumination light is not limiting the rate of the process, and a rise in temperature of $10°$ may double the rate of the reaction $(Q_{10} = 2)$. Under low illumination a similar rise in temperature makes little difference to the rate of photosynthesis. This indicates that there is a distinct purely chemical stage in photosynthesis in addition to a photochemical stage.

6. Estimation of the percentage of carbon dioxide and oxygen in a gas sample (Eggleston's method)

The apparatus is a simple capillary tube to which is attached a brass screw; this should be carefully greased to prevent a leak (fig. A.5). Turning the screw allows for a gas sample to be taken in or out of the tube.

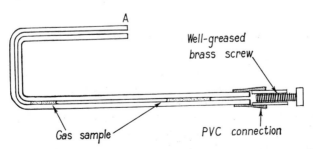

Fig. A.5. Eggleston's capillary device for gas analysis. (Note: the long arm of the capillary should be 30 cm. in length.)

First turn the screw fully in, dip the end of the capillary (A) in water and unscrew until about 5 cm length of water has been taken up. To take up a gas sample, adjust the water column in the capillary so that it reaches the open end, then place the end in the atmosphere to be sampled and unscrew until about a 10 cm length of gas has been drawn in. Seal this by drawing in a little more water. Place the tube on its side in a water-bath for about ten minutes to bring it to a uniform temperature. Taking care not to warm the capillary tube with your hands, measure the length of the gas sample. Then place the open end in a strong solution of potassium hydroxide (*caution:* take care with this substance which is highly caustic and can cause burns and damage clothing) to absorb the carbon dioxide. Screw in until the gas bubble is almost at the end of the capillary. Unscrew and draw in a little of the reagent. Using the screw, shunt the gas sample backwards and forwards to allow it to come into contact with the potassium hydroxide. Leave to stand for five minutes and then measure the length of the bubble as before. Any shortening indicates how much carbon dioxide was present.

The last stage is then repeated using alkaline pyrogallol* (*caution;* take care with this reagent which is highly caustic and can cause burns and damage clothing) to absorb the oxygen. It is always necessary to estimate the carbon dioxide separately, as the pyrogallol absorbs both oxygen and carbon dioxide. At the end of each run rinse the tubes in dilute acid to remove all traces of the alkaline reagent.

(Adapted from *Biology Students' text*, Year III, of the Nuffield Foundation Science Teaching project. Longmans/Penguins, 1966.)

7. General technique of chromatography

Chromatography is used for separating and identifying unknown mixtures of closely related organic and inorganic substances.

Most chromatographic apparatus consists of some solid medium or *adsorbing substance*. This is arranged over a solvent, usually composed of two or more substances, which can pass through the adsorbent. A mixture of the substances to be separated is placed on the adsorbent so that, as the solvent passes through them, they may be drawn along too. Separation of the substances in the mixture takes place if some of them are more strongly adsorbed than others. The *partition effect* is usually also important as the substances will assort themselves on the chromatogram according to their partition coefficients in the solvents.

Ideally the position reached by any particular substance should always be the same under identical conditions, and so it is possible to identify the substances formed. In practice though a good deal of care is needed if reproducible results are to be obtained.

A number of different chromatographic techniques are useful in biology. The simplest is perhaps *filter-paper chromatography*, using long strips or large sheets. There is also *thin-layer chromatography* which uses thin layers of silica-gel mixed with plaster of Paris on glass plates. These two techniques are useful for identifying mixtures of substances. Filter-paper techniques are probably better when plenty of time is available, but thin-layer methods are particularly useful as good separations can be obtained in under an hour. *Column chromatography* is used to separate mixtures in solution and is carried out in a tall glass column filled with cellulose or aluminium oxide.

Each of these techniques is described below; a full table of substances, methods of extraction from plant material, solvents, developing sprays and R_f values is given on p. 212. Caution is needed in handling many of these chromatograms so handle materials with rubber gloves and use the fume cupboard.

8. Filter-paper chromatography using long strips

The chromatogram jar (see fig. 3.16)
Fit a tall narrow glass jar (about 45 cm high) with a cork that has been bored to take a glass rod fitted with a small hook at its end. This glass rod should be able to slip up and down in the cork. Place 2 cm of the solvent (see p. 212) in the jar and cut a long strip of Whatman No. 1 filter-paper (it can be obtained in a roll of the required width). Solvents run at rates which change with different papers and according to the way the paper is hung. The box containing the filter-paper is normally marked to indicate the way the solvent should flow. Arrange the glass rod and filter-paper strip so that the filter paper can hang freely in the air without touching the solvent.

Loading the paper

Mark in pencil a spot 3 cm from the base of the paper strip and load the unknown, or known, comparison substance carefully on to the spot, using a fine capillary or fine wire loop. It is important to load plenty of material (unless it is a concentrated solution of a single known substance), but it is also important to keep the spot less than 5 mm diameter.

Running the chromatogram

Attach the loaded filter-paper to the hook by means of a pin or paper-clip. Place it in the jar for ten minutes for the atmosphere inside the jar to equilibrate. Then lower the glass rod so that the paper dips 5 mm into the solvent. Make sure that the cork is a proper seal, if necessary smear a little Vaseline around it and the rod. The solvent will rise slowly up the paper, carrying the material from the spot with it. When the solvent has risen about 30 cm up the paper it can be removed. Mark the height that the solvent has reached (the solvent front).

Development of the chromatogram

Dry the paper carefully, *in a fume cupboard*. Remember that many solvents are inflammable or noxious. Some spots will be coloured already, others need a developing spray. Spray the chromatogram in the fume cupboard taking care not to inhale the spray vapour. Mark the spots as they appear and make a note of their colour as they often fade quickly. It is also worth while to view the chromatogram in ultra-violet light (see p. 221).

Identification of the products

Identification of the spots is possible by running a parallel chromatogram alongside that with the unknown substances. This chromatogram is loaded with

The table is constructed in 3 parts: Part A—solvents; Part B—extraction and development; Part C—(p. 214 and 215) R_f values

Part A

Substances	Solvents
Chlorophyll and Carotenoid pigments	*Using thin-layer or filter-paper chromatography* 100 parts petroleum ether (B.P. 100°–120°C.), 12 parts pure propanone (acetone). The solvent listed below gives a better separation with thin-layer chromatograms. *Using thin-layer or column chromatography* (for rest of details of column chromatography see p. 219) 100 parts petroleum ether (B.P. 60°–80°C.), 20 parts pure propanone (acetone).
Plant acids	100 parts butyl methanoate, 40 parts 98 per cent methanoic acid, 10 parts distilled water. *Take care with these substances.* Add 0.5 g sodium methanoate to each 100 cm³ of solvent and sufficient solid bromophenol blue to turn the solvent a pale orange colour.
Amino-acids Sugars and Plant acids	Phenol-water. Take 28 g of pure crystalline phenol, which should be white and unoxidized. *Take care when handling this substance; it is preferable to use rubber gloves.* Place the phenol in a stoppered separating funnel and add 12 cm³ of distilled water. This is the minimum quantity for work with a filter-paper strip, use a larger quantity if necessary. Add a little sodium chloride and shake thoroughly. Fill the funnel with nitrogen to prevent oxidation and leave until the layers separate out, which usually takes at least an hour. The lower layer is the phenol saturated with water; reject the upper layer.
Amino-acids and Sugars	120 parts butan-1-ol, 30 parts glacial ethanoic (acetic) acid, 60 parts distilled water.
Amino-acids	*Using two-way filter-paper chromatography* This system gives a more effective separation. After running the chromatogram in either of the above two solvents, dry it but do not develop it. Turn the paper on edge and run it in the following solvent: 180 parts ethanol, 10 parts ammonium hydroxide, 10 parts distilled water.
Anthocyanidin pigments	40 parts butan-1-ol, 10 parts glacial ethanoic (acetic) acid, 50 parts distilled water. Make up the solvent in a separating funnel. The lower layer is the required solvent, reject the upper layer.

Table of Chromatographic Techniques (*continued*)
Part B

Substances	Methods of Extraction from Plant material	Development
Chlorophyll and Carotenoid pigments	Grind in pure propanone (acetone) or as described on p. 219	No development necessary Ultra-violet light is particularly useful for viewing spots.
	,,	,,
Plant acids	Grind in 70 per cent ethanol	Dry the chromatogram carefully, preferably in a fume cupboard. Sometimes further development is worthwhile by holding the chromatogram over a bottle of ammonium hydroxide. Do not let the chromatogram get too blue. The spots sometimes become clearer if the chromatogram is left for some days.
Amino-acids	Grind in water or 80 per cent ethanol	Dry the chromatogram carefully, preferably in a fume cupboard as phenol fumes are noxious. Spray with 2 per cent ninhydrin in butan-1-ol. (Care: ninhydrin is a carcinogen.) Heat the chromatogram strongly for the spots to appear. Aerosol can sprays may be used.
Sugars	Grind in 50 per cent ethanol	Dry the chromatogram carefully, preferably in a fume cupboard as phenol fumes are noxious. Spray with either a 3 per cent solution of *para*-anisidine hydrochloride in butan-1-ol plus a few drops hydrochloric acid or with 10 per cent solution of resorcinol in propanone (acetone) plus a few drops hydrochloric acid.
Plant acids	Grind in water	Dry the chromatogram carefully, preferably in a fume cupboard as phenol fumes are noxious. Spray with bromothymol blue adjusted to pH 8.5 with sodium hydroxide.
Amino-acids	Grind in water or 80 per cent ethanol	Dry the chromatogram carefully (butan-1-ol is highly inflammable) spray with 2 per cent ninhydrin in butan-2-ol. (Care: ninhydrin is a carcinogen.) Heat the chromatogram strongly for the spots to appear.
Sugars	Grind in 0.1 M sodium ethanoate in water	Dry the chromatogram carefully (butan-1-ol is highly inflammable) spray with either a 3 per cent solution of *para*-anisidine hydrochloride in butan-1-ol plus a few drops hydrochloric acid. Heat the chromatogram strongly for the spots to appear.
Amino-acids	(as above)	Dry the chromatogram carefully (ethanol is highly inflammable). Spray with 2 per cent ninhydrin (care: ninhydrin is a carcinogen) in butan-1-ol. Heat the chromatogram strongly for the spots to appear.
Anthocyanidin pigments	Grind in 80 per cent ethanol	No development necessary. Treatment with ammonia vapour makes the spots blue and sometimes intensifies them.

Table of Chromatographic Techniques (*continued*)
Part C
Table of R_f Values arranged by solvent system as in Part A

| | Identification of Spots | | | | | |
| | Thin-layer Chromatography | | | Filter-paper Chromatography | | |
Substances	Substance	R_f value	Colour	Substance	R_f value	Colour
Chlorophyll and Carotenoid pigments	Chlorophyll b	0.10	Yellow-green	Chlorophyll b	0.45	Green
	Chlorophyll a	0.11	Blue-green	Chlorophyll a	0.65	Blue-green
	Xanthophyll	0.22	Yellow	Xanthophyll	0.71	Yellow-brown
	Phaeophytin	0.28	Grey	Phaeophytin	0.83	Grey
	Carotene	0.90	Yellow	Carotene	0.95	Yellow
	Chlorophyll b	0.17	Yellow-green			
	Xanthophyll (?)	0.19	Yellow			
	Chlorophyll a	0.23	Blue-green			
	Xanthophyll	0.35	Yellow			
	Phaeophytin	0.44	Grey			
	Carotene	0.96	Yellow			
Plant acids	2,3-dihydroxybutanedioic acid (Tartaric acid)	0.32	Yellow against a purple background	2,3-dihydroxybutanedioic acid (Tartaric acid)	0.20	Yellow against a purple background
	2-hydroxypropane-1,2,3-tricarboxylic acid (citric acid)	0.50		2-hydroxypropane-1,2,3-tricarboxylic acid (citric acid)	0.25	
	2-hydroxybutanedioic acid (malic acid)	0.61		Ethanedioic acid-2-water (oxalic acid)	0.32	
	2-oxopropanoic acid (pyruvic acid)	0.85		2-hydroxybutanedioic acid (malic acid)	0.37	
	Butanedioic acid (succinic acid)	0.92		Butanedioic acid (succinic acid)	0.57	
Amino-acids	Glycine	0.29	Orange-red	Glutamic acid	0.38	Orange-red
	Arginine	0.32	Deep red-purple	Glycine	0.50	Brown-purple
	Cystine	0.32	Pink	Tyrosine	0.66	Deep purple
	Valine	0.41	Red-purple	Arginine	0.70	Red-purple
	Phenylalanine	0.52	Brown	Alanine	0.72	Blue-purple
				Leucine	0.91	Deep purple
				Proline	0.95	Yellow
Sugars				Glucose	0.31	Red with resorcinol, yellow-brown with anisidine
				Sucrose	0.35	
				Fructose	0.46	

Table of Chromatographic Techniques (*continued*)
Part C (*continued*)
Table of R_f Values arranged by solvent system as in Part A

Identification of Spots

Substances	Thin-layer Chromatography			Filter-paper Chromatography		
	Substance	R_f value	Colour	Substance	R_f value	Colour
Plant acids				2,3-dihydroxybutanedioic acid (Tartaric acid)	0.23	Yellow against a blue background
				2-hydroxypropane-1,2,3-tricarboxylic acid (citric acid)	0.32	
				Ethanedioic acid-2-water (oxalic acid)	0.35	
				2-hydroxybutanedioic acid (malic acid)	0.43	
				2-oxopropanoic acid (pyruvic acid)	0.59	
				Butanedioic acid (succinic acid)	0.60	
Amino-acids	Glycine	0.20	Orange-red			
	Arginine	0.23	Deep red-purple			
	Tyrosine	0.25	Purple			
	Cystine	0.26	Pink			
	Valine	0.39	Red-purple			
Sugars	Sucrose	0.24	Yellow-brown			
	Fructose	0.31	Yellow-brown			
	Glucose	0.43	Yellow-brown			
Amino-acids	—			—		
Anthocyanidin pigments				Delphinidin	0.59	Blue-purple
				Pelargonidin	0.73	Bright red
				Peonidin	0.74	Magenta
				Cyanidin	0.79	Mauve-purple

a known substance that is thought to be present in the unknown mixture. The resulting chromatograms are then compared and the R_f values of the spots calculated:

$$R_f = \frac{\text{Distance moved by solute spot}}{\text{Distance moved by solvent front}}$$

In addition, R_f values can be found in tables, but it must be emphasized that it is almost always necessary to run a comparison chromatogram with a known substance before a certain identification can be made.

It is also possible to make a rough assessment of the comparative amounts of the substances in the mixture by comparing the size and density of the various spots.

9. Filter-paper chromatography using large sheets

Fit a large glass jar (e.g. an old battery jar) with a cover as shown in fig. A.6. Arrange the glass rod so that the filter-paper sheet can hang to the bottom of the jar. Load the substances on to the paper as described above. Put about 2 cm of

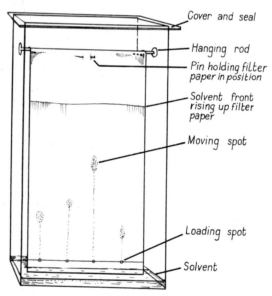

Fig. A.6. The chromatogram jar.

solvent in the bottom of the jar and also stick a few filter-papers, moistened with solvent, to the side of the jar and dipping in the solvent. This helps to keep the atmosphere inside the tank saturated with solvent vapour. Lower the paper in carefully and fix it to the glass rod with a clip or pin small pieces of PVC tubing can be used to fix the glass rod in position. Seal the jar as quickly as possible after smearing Vaseline around the rim or by using a strip of Plasticine around the edge.

Comparison of known and unknown spots is much easier than with the thin strips described above.

If a square tank is used, *two-way chromatograms* can be made by running the chromatogram as described and then, after drying the paper, using a new solvent and turning the paper on edge (see p. 212).

10. Thin-layer chromatography

Specially prepared celluloid film with thin-layer coating is available commercially but it is also possible to prepare your own plates:

Preparation of the glass plates

Old half-plate photographic plates, carefully cleaned in chromic acid, make very suitable thin-layer plates. Arrange about twelve of these along a flat bench as shown in fig. A.7.

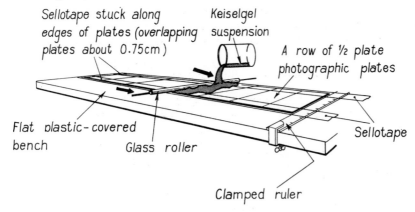

Fig. A.7. Spreading thin-layer chromatogram plates. (From Baron, W. M. M. (1964). *School Science Review*, No. 158, p. 62.)

The plates should be arranged side-by-side with their long sides touching. Clamp a ruler at one end so that the plates cannot slip. Then stick Sellotape along the two outer edges of the plates. This prevents the plates from slipping and also keeps the roller at exactly the right height above the plates.

Take 30 g of the adsorbant (Kieselgel G after Stahl), and add it to 60 cm³ of distilled water, stirring quickly and thoroughly. Pour the Kieselgel on to one end of the plates and spread it with a glass roller, moving the roller steadily and evenly across the plates. *Never move it more than once.* Take care to keep sufficient Kieselgel in front of the roller. Leave the plates to dry for one hour, then remove the Sellotape and incubate them at between 80° and 100°C. for one hour. The plates may then be kept in a desiccator for a few days until they are needed.

Before the plates are used clean off any traces of Sellotape and also scrape a 1 cm band of Kieselgel from the vertical edges of the plate (see fig. A.8), this prevents 'billowing' of the solvent as it comes into contact with the edges of the plate. Handle the plate at the edges only, never touch the Kieselgel.

Loading the chromatogram

Spots of known or unknown substances should be applied about 10 mm from the lower end of the Kieselgel. Spots *must be small*, 2–3 mm is quite large enough. Use a minute wire loop for loading, as this causes less damage to the Kieselgel than a capillary. Only one application of concentrated known substances is usually needed, but considerably more material of unknown and usually dilute mixtures will have to be added. Use scratched marks at the top of the plate to label the spots.

Running the chromatogram

Sufficient solvent is prepared to fill the chromatogram jar about 15 mm deep. In this way it will come into proper contact with the Kieselgel without washing

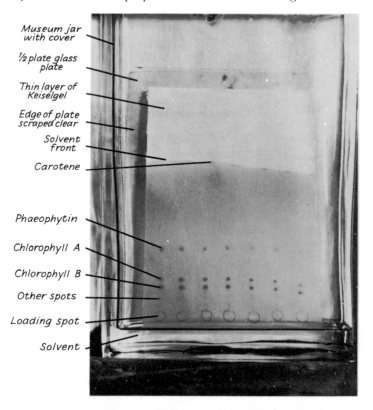

Fig. A.8. A thin-layer chromatogram plate separating the chlorophyll pigments. (From Baron, W. M. M. (1964). *School Science Review*, No. 158, p. 62.)

away the loaded spots. The most suitable jars are probably the small glass jars, complete with ground glass covers that are often used for museum specimens. These will take two plates at a time and about 60 cm³ of solvent is usually sufficient. After placing the solvent in the jar keep the top on for a few minutes to allow the solvent and its vapour to come into equilibrium. Added saturation of the tank atmosphere can be obtained by coating the sides of the tank with filter-paper, dipping in the solvent. Then carefully put the plate in and replace the cover and seal the jar with a Plasticine strip or Vaseline. Ether-containing solvents usually take about twenty-five minutes to run. Phenol will take about an hour. Remove the plate when the solvent front has risen to 10 cm.

Identification of the products

The spots may be identified from R_f values; the main problem is, however, that owing to lack of uniformness in the Kieselgel layer, exactly comparable R_f values are seldom obtained. This can be overcome by using known substances for comparison or by comparing with a standard dye range. A mixture of two dyes (Sudan Red G and Indophenol) can be used as a control with some solvents.

11. Extraction of the photosynthetic pigments (see p. 56)

Materials. Leaves of the nettle (*Urtica dioica*).
Method. Place a few leaves in a mortar with about 30 cm³ of pure propanone (acetone) and grind to extract the pigments. Filter the extract through glass wool, using a Büchner funnel, and place the filtrate in a separating funnel. Add an equal volume of petroleum ether (B.P. 100°–120°C.) and shake the mixture. Wash the extract three times with distilled water, rejecting the watery layer each time. Add solid sodium sulphate to help break the emulsion down and let the solution stand for some minutes over sodium sulphate, it is then ready for use.

12. Separation of the chlorophyll pigments by column chromatography (see p. 56)

Material. Freshly made extract of the pigments.
Method. Fill a 30 cm, 1 cm bore glass tube with pure Whatman cellulose powder. (Aluminium oxide usually gives a quicker separation but sometimes some of the pigments break down.) A small piece of glass wool is used as a pad at the lower end (see fig. 3.17). It is important to take care with this loading, only a little cellulose being added at a time and a glass ramrod being used to firm it between each addition. Then wash the column through with a mixture of 100 parts petroleum ether (B.P. 60°–80°C.) to 20 parts pure propanone, using a filter pump to help draw the mixture through; the column is then ready for use. Add 5 cm³ of the chlorophyll extract to the top of the column and allow it to sink into the cellulose. Set up a dropping funnel filled with solvent over the column, adjust it to drip steadily and the pigments will be washed down the column. The pigments can be seen to separate in the column as they are washed down, the least adsorbed substances travelling most quickly. After a few minutes the carotene solution will begin to drip out of the column; collect it and the remaining pigments (in the order: carotene, phaeophytin, xanthophyll, chlorophyll *a* and chlorophyll *b*) in a series of test-tubes held under the column. (Adapted from Baron (1960). *School Science Review*, No. 145, p. 93.)

13. Setting-up and calibration of the spectrometer (see p. 59)

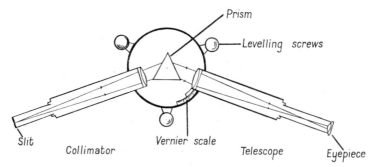

Fig. A.9. The spectrometer.

If a wavelength or constant-deviation spectrometer is not available the ordinary spectrometer can be calibrated so that the wavelengths of absorption and emission bands can be measured.

It is first necessary to set up the apparatus so as to obtain maximum resolution and the desired dispersion of the spectrum. First focus the cross-lines by adjusting the telescope eyepiece (see fig. A.9), then focus the telescope on a distant object. Bring the telescope into line with the collimator and illuminate the latter. Adjust the position of the collimator slit in or out until a sharp image of the slit is formed in the plane of the eyepiece cross-lines. Place the prism as shown in the diagram (fig. A.9), clamp it down and place a black cloth over the central part of the apparatus. By rotating the eyepiece telescope different parts of the spectrum can be viewed and the angles on the vernier scale noted.

The light produced by a mercury vacuum tube or arc is most useful for calibration, as it produces a series of sharp emission bands, the wavelengths of which are known and can be compared with the angle on the vernier scale.

MERCURY VAPOUR SPECTRUM	
Wavelength (nm)	*Colour*
576.96	Yellow
579.07	Yellow
546.07	Green
435.82	Violet

Fainter lines are also visible at: 623.2, 615.2, 495.97, 491.64, 407.81, 404.68, 365.0, 313.1 and 312.6 nm.

Plot the angles and wavelengths on a graph so that any intermediate wavelengths may subsequently be determined from the curve obtained.

Measurement of the wavelengths of absorption and emission spectra is much simpler using the Hilger's wavelength spectrometer. Here the collimator and telescope are fixed, and as the prism is rotated the different parts of the spectrum are visible and their respective wavelengths can be read off directly on the drum. It is, however, wise to check the apparatus before use, and this is most simply done by examining the sodium flame, which has two close orange emission bands at 589.0 and 589.59 nm.

14. Investigation of the fluorescence of the chlorophyll pigments
(see p. 63)

Materials. Fresh leaves of lesser celandine (*Ranunculus ficaria*), fresh or dried leaves of nettle (*Urtica dioica*).
Method. Grind up a small quantity of the leaves in propanone (acetone) and centrifuge the extract. Examine in ultra-violet light; a strong red fluorescence should be seen.

A note on the ultra-violet light source
A useful ultra-violet lamp can be constructed quite easily. It consists of a 125 watt special blue bulb, but requires a choke, capacitor and special three-contact lampholder. The wavelengths of ultra-violet light emitted by this lamp should not cause damage to the retina of the eye, during short exposures, but, on the other hand, the lamp cannot be used for any sterilization procedures. The lamp should be allowed to warm up for several minutes before use. A thick glass screen should be placed between the observer and the lamp as, on occasion, these ultra-violet lamps have been known to explode.

15. Investigation of the catalytic properties of extracted chloroplasts (the Hill reaction) (see p. 62)

Material. Fresh leaves of spinach.
Method. Prepare 40 cm^3 sucrose buffer (see p. 256). Cool this by immersing the flask in a freezing mixture and at the same time cool a pestle and mortar. Chop up some fresh spinach leaves into the mortar and then grind them to extract the chloroplasts in about 10 cm^3 of ice-cold buffer. Centrifuge at medium speed for three minutes to throw down cell-wall detritus and starch grains. The centrifuge may be arranged so that the large tubes contain freezing mixture, and these hold smaller, inner tubes containing the chloroplast extract. Decant the chloroplast suspension into clean centrifuge tubes and recentrifuge at high speed for ten minutes. Reject the supernatant and resuspend the chloroplasts in 4 cm^3 ice-cold buffer.

1,2	1 cm^3 very dilute aqueous 2,6-dichlorophenol-indophenol	0.5 cm^3, 0.5 M-KCl	0.5 cm^3 chloroplast suspension
3,4	1 cm^3 very dilute aqueous 2,6-dichlorophenol-indophenol	0.5 cm^3, 0.5 M-KCl	0.5 cm^3 boiled chloroplast suspension (control)
5,6	no dye (control) 1 cm^3 distilled water	0.5 cm^3, 0.5 M-KCl	0.5 cm^3 chloroplast suspension

Add the purified chloroplast extract to each tube, except 3 and 4, to which are added boiled chloroplast extract. Illuminate tubes 1, 3 and 5 brightly and leave the other tubes in the dark. Take the time for the dye to bleach in the light-treated tubes. Compare the colours with those kept in the dark. What conclusions can you draw about the reducing properties of illuminated chloroplasts?

16. Investigation of the uptake of carbon dioxide into the leaf using ^{14}C radioactive tracer (see p. 48 and 51)

Material. Small, actively growing shoots of many plants such as tomato or *Zebrina* (*Tradescantia*). Variegated plants could also be used. These shoots need not be rooted.

If translocation is to be investigated it is necessary to modify the apparatus so that only single leaves of the intact plant are treated with tracer.

Method

1. *Apparatus.* Set up the apparatus shown in fig. A.10. This consists of a dropping funnel about 10 cm long, fitted with a rubber bung, tube and tap. The funnel is arranged to contain two specimen tubes resting on a piece of pierced expanded polystyrene. The smaller tube contains the tracer solution; the larger contains water and the shoot, some leaves of which should be covered with black polythene, others with transparent polythene, using a paper clip.

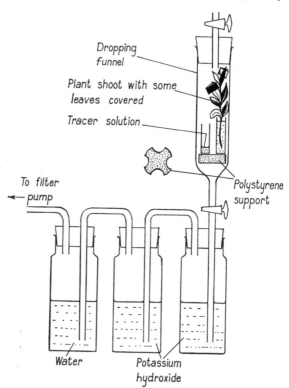

Fig. A.10. Apparatus used for investigating the uptake of $^{14}CO_2$ into the plant (see p. 51).

The dropping tube leads directly into a wash-bottle containing a strong solution of potassium hydroxide. This is connected to a second similar bottle and finally to a third which should contain water. At the end of the experiment the tube from this bottle is connected to a filter pump.

A 100 watt bulb is arranged about 30 cm from the tube, with a sheet of glass in between to prevent heating, and the whole apparatus is placed in a fume cupboard. The bench should be covered with a sheet of tarred paper in case of spills. Make sure that there are no leaks in the apparatus.

2. *Tracer technique.* When the apparatus is set up, carefully remove the calculated quantity of the solution of radioactive tracer from the phial or serum bottle and place it in the smaller specimen tube. This will probably be only a few drops. Use a disposable syringe or special pipette fitted with a rubber bulb; *under no circumstances use a mouth pipette.* Wear rubber gloves throughout the experiment. Carry out this operation on an enamel tray in the fume cupboard. The ^{14}C is usually obtained in the form of a solution of sodium carbonate and the quantity required must be calculated so as to give a strength of $10\,\mu Ci$ (microcuries) in the apparatus. ^{14}C is a weak β-emitter ($0.155\,MeV$) and needs a sensitive Geiger–Muller (GM) tube for proper detection and counting (e.g. Mullard MX 168 or, better, MX 168/01).

Place the specimen tube in a deep freeze or freezing mixture until the carbonate solution is frozen, taking precautions to ensure that the tube is not knocked over. Note that solutions of radioactive sodium carbonate should not be left for long exposed to the atmosphere, as gaseous exchange takes place; accordingly keep all such solutions properly stoppered.

When the carbonate is frozen place a few crystals of solid sodium bisulphate on the surface of the carbonate. Place the tube in the dropping funnel, put in the rubber bung and close the top tap, but leave the lower open. The filter pump should *not* be on at this stage. Turn on the lamp; as the carbonate thaws it reacts with the bisulphate and the radioactive carbon dioxide is evolved. After half an hour close the lower tap and leave the apparatus with the light on for between 12 and 24 hours. Finally, open the lower tap, turn on the filter pump and then open the upper tap. Draw air through the apparatus for at least half an hour to remove the radioactive carbon dioxide. Then carefully remove the rubber bung and check for radiation; if necessary continue to draw air through the apparatus for a further period. When there is little carbon dioxide left, take out the plant material and cut off the wet parts of the shoot, put these in the waste bottle (see below).

3. *Autoradiographic technique.* This is used to investigate the sites of absorption of the radioactive carbon dioxide. Place the shoots on a piece of paper, 16×12 cm, carefully separate out the leaves (still using rubber gloves) and stick the shoots down to the paper, using one or two pieces of sticky tape. Make a note to show where the leaves were covered and remove the clips and covers. Place the paper in a thin plastic bag to prevent any radioactive plant juices contaminating the film or holder. Take the material to the darkroom and place it in a special half-plate X-ray exposure film holder. Arrange the material so that the emulsion surface of the X-ray film is facing the plant. There will then be a sheet of plastic and paper between the film and the plant but no pieces of tape. Close up the film holder and leave for between 36 and 48 hours. Do not press the material too hard in the folder if the shoots are full of sap.

Suitable X-ray films are Kodak *Kodirex* and *Crystallex.* The former is the faster and is probably the more satisfactory film, though the latter gives a very clear print. Deep-red safe lights may be used for loading and development, though Kodak recommend *Wratten* 6B (brown). After exposure remove the film from the folder and develop it in Kodak D-19b for 5–12 minutes at 20°C. Fix in Kodak *Unifix* or similar fixer.

Compare the autoradiographs with the sketches showing which leaves were covered. Mark the autoradiographs accordingly and make what deductions you can about the sites of absorption of the carbon dioxide.

4. *Identification of the substances formed using autoradiography of chromatograms.* Remove the plant shoots from the paper backing and place them in a beaker containing a small quantity of 90 per cent ethanol. Use the minimum quantity to cover the plant. Cover with a watch glass and heat the beaker on a water-bath in a fume cupboard for ten minutes.

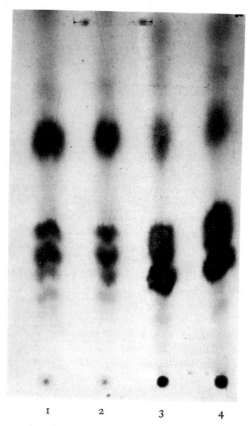

I   2   3   4

Fig. A.11. Autoradiograph of chromatogram of extracts of tomato (1 and 2) and *Zebrina* (3 and 4) after the plants had been allowed to photosynthesise in an atmosphere containing $^{14}CO_2$. (Chromatographic solvent: butan-1-ol, ethanoic acid water. Fourteen days' exposure on Kodirex.)

Spot the solution on to three or more sheets of filter paper for chromatography (see p. 211). Run the chromatograms in butan-1-ol or other solvents. Spray with appropriate reagents to identify amino-acids, sugars and plant acids. After marking and identifying the spots make a rough estimate of the distribution of tracer, using a scaler, and make a note of the readings from the various spots. Finally, cover the chromatograms with plastic (both sides) and set them up in the dark-room with X-ray film in the usual way. Leave for 14 days and then develop the films. Compare the autoradiographs and chromatograms carefully and deduce which substances have accumulated tracer from the radioactive carbon dioxide. An example of such an autoradiograph is illustrated in fig. A.11.

5. *Washing and disposal of waste.* ^{14}C has a long half-life (5568 years), so special care is necessary in handling it and in the disposal of waste. At the conclusion of the experiment wash the glassware, dropping funnel and rubber bung with dilute potassium hydroxide, transferring the washings and all the radioactive liquids (from the wash bottles and specimen tubes) to a specially-labelled, large-stoppered waste jar. A second washing of the contaminated glassware should be carried out, using a little warm brine and detergent; these washings should also be regarded as 'hot' and put into the waste jar. Once the contamination has been removed the glassware may have a final wash in warm water and detergent. Dry the apparatus with tissue and keep it solely for this work. The use of special warning tape is useful here and throughout the experiment. Any solid wastes (e.g. plant material) should be kept in a separate bottle. Regulations for the disposal of wastes are strict. Full details of various techniques are given in Faires, R. A., and Parkes, B. H. (1958) *Radioisotope Laboratory techniques*; Newnes, London. The reader in the United Kingdom is also advised to consult the Administrative Memorandum 1/65 of the Dept. of Education and Science, Curzon Street, London, W.1, which gives further details and regulations for the use of radioactive materials in schools. Permission to use radioactive materials in schools must also first be obtained from the department. Radioactive materials may be obtained from The Radiochemical Centre, Amersham, Bucks.

17. Investigation of the synthesis of starch from glucose in the dark by leaves of Pelargonium (see p. 65)

Material. Young leaves of *Pelargonium*, or tobacco.
Method. Sterilize two small, but deep Pyrex dishes together with suitable glass covers. Sterilize 20 cm³, 5 per cent glucose solution and pour it into one dish, pour 20 cm³ sterile distilled water into the other (as a control). Using a cork borer, cut several 1 cm diameter discs from a leaf of a *Pelargonium* plant that has been in the dark for at least twenty-four hours and sterilize them by shaking for three minutes in 1 per cent sodium hypochlorite solution. Wash them in sterile distilled water and float one in each dish. Label the dishes and leave them in the dark for three days. Finally, take them out and mark the leaf disc that has been in the glucose by cutting out a small nick. Remove the chlorophyll by boiling in 90 per cent alcohol for some minutes and test with iodine for starch formation.

APPENDIX TO CHAPTER 4

1. Investigation of the rate of respiration by the Pettenkofer technique
2. Determination of the respiratory quotient (RQ) by means of the Ganong respirometer

3. The Warburg manometric technique
4. Determination of the respiratory quotient by Warburg manometry
5. Investigation of the effect of addition of an inhibitor on the respiration rate
6. Calculation of the flask constants of the Warburg manometer
7. Identification of acetaldehyde as an intermediate in anaerobic respiration
8. Identification of some plant acids by simple chemical tests
9. Identification of plant acids by chromatography
10. Investigation of polyphenol oxidase as an example of a terminal oxidase
11. Investigation of the dehydrogenase activity of etiolated pea shoots
12. Spectroscopic examination of the cytochromes

1. Investigation of the rate of respiration by the Pettenkofer technique (see p. 71)

Materials. Germinating seeds at various stages, ripening fruits, etc.
Method. Apparatus.

Set up the apparatus shown in the diagram fig. 4.2, this consists of seven main pieces:

1. A small piston-action blowing motor.
2. A tower containing moist soda-lime to absorb carbon dioxide. (Soda-lime is a mixture of sodium hydroxide and calcium oxide.)
3. A pressure control unit consisting of a capillary tube which can slide up and down inside a boiling-tube. Raising or lowering the capillary tube will control the rate of flow of gas through the apparatus.
4. A check-flask which contains calcium hydroxide; if this goes cloudy the soda-lime tower must be renewed immediately.
5. A respiration chamber, the size of which will depend on the material being investigated. For small amounts a bottle or gas jar can be used, for larger specimens an old battery jar can be satisfactorily modified. The chamber should be covered in black paper to prevent photosynthesis taking place.
6. The Pettenkofer tubes; these are long horizontal tubes containing $25\,cm^3$ 0.05 M-barium hydroxide (baryta) together with $25\,cm^3$ boiled distilled water. A two-way tap is fitted to facilitate easy changing of tubes.
7. Finally, check flasks containing calcium hydroxide to make sure that all the respiratory carbon dioxide has been removed in the Pettenkofer tube.

In setting up the apparatus it is important to use good-quality tubes and rubber bungs, otherwise leaks may result.

Procedure. Weigh out the respiring material and place it in the respiration chamber, test the apparatus for leaks and regulate the flow so as to allow about 100 bubbles per minute through the Pettenkofer tube (filled with $50\,cm^3$ distilled water). Meanwhile fill the other Pettenkofer tube with $25\,cm^3$ fresh 0.05 M-baryta and $25\,cm^3$ boiled distilled water. Cork-up, change the two-way tap, adjust the rate of bubbling and leave for between one and four hours. At the end of this time remove the Pettenkofer tube and pour its contents carefully into a conical flask. Wash the tube with boiled distilled water and add the washings to the flask.

Titration. The baryta from the Pettenkofer tube should be titrated immediately against 0.1 M-HCl using phenolphthalein as indicator. During the experiment the carbon dioxide has reacted with the baryta to form insoluble barium carbonate. In the titration the HCl reacts with the unchanged baryta only.

Amount 0.1 M-HCl required $= x \, cm^3$
Time in hours $= y \, hours$

∴ Amount 0.05 M baryta that reacted with the CO_2 in one hour $= \dfrac{25 - x}{y} \, cm^3$

Thus the amount of CO_2 absorbed in one hour $= \dfrac{(25 - x)}{y} \times \dfrac{11.2}{10} \, cm^3$ at N.T.P.

Knowing the weight of the respiring material, the carbon dioxide production per hour per gm can be calculated.

During the running of the experiment the normality of the baryta should be checked by titration with 0.1 M-HCl and if necessary, allowance made in the calculation of the results. Accuracy depends to a large extent on efficient handling of the baryta which must be kept in a flask stoppered with a soda-lime tower, the baryta being removed from a tap at the bottom of the flask.

2. Determination of the respiratory quotient (RQ) by means of the Ganong respirometer (see p. 74)

Material. Germinating peas.
Method. Set up two sets of apparatus as shown in fig. 4.5. Fill one apparatus with a strong solution of potassium hydroxide, enclosed about 5 g of peas in the inverted U-tube, cork-up (if necessary, wax the cork), adjust the level and record that in the graduated tube. As there will be a rise of liquid in this tube, make sure that this starting level is near the bottom of the graduated tube.

Set up a similar apparatus with the same weight of germinating peas but having water or paraffin in the system, in this case arrange for the level of liquid to be half-way up the graduated tube. Record the level and leave both sets of apparatus for twenty-four hours. If really accurate results are to be obtained, a third, control respirometer should also be set up. This should contain water and dead, sterile peas. If it shows any change, adjustments must be made to the readings of the other respirometers.

The amount of oxygen consumed is given by the rise of liquid in the apparatus containing the potassium hydroxide (which absorbs carbon dioxide as soon as it is produced). The amount of carbon dioxide produced is found by comparing the volume changes in both sets of apparatus. If more carbon dioxide is produced than oxygen utilized, then there will be an expansion in the apparatus containing water, and this expansion must be added to the oxygen value to give the amount of carbon dioxide produced. If there is no change in level, then the amount of oxygen utilized is the same as the amount of carbon dioxide released (and the $RQ = 1$). If there is a reduction in volume, then less carbon dioxide is produced than oxygen utilized, and the amount of carbon dioxide produced is given by the oxygen value less this reduction in volume. From these values the RQ of the germinating peas can be calculated (see also p. 73):

$$RQ = \frac{\text{Volume carbon dioxide produced}}{\text{Volume oxygen utilized}}$$

3. The Warburg manometric technique (see p. 71)

Principle

In this technique the respiring material is enclosed in the main chamber of a small flask (see figs. 4.3 and 4.4) and kept at a constant temperature in a thermostatically controlled water-bath usually at 25° or 30°C. A small manometer, filled with Brodie's fluid* is attached to the flask and registers any changes in volume due to gases being taken up or produced. In practice, the volume is kept constant by adjustment to the reservoir at the base of the manometer and readings are taken as heights of liquid in the open side of the manometer tube that are required to keep the volume of the flask system unchanged. In many investigations of the respiratory rate the amount of oxygen utilized is found by placing a few drops of strong potassium hydroxide in the centre well, to absorb carbon dioxode. The amount of oxygen taken up in a particular time is given in terms of height changes in the open manometer arm. If these values are to be corrected to mm³, then the flask constants must be applied (see exp. 6, below).

The manometer flask is also provided with a side arm from which inhibitors or special substrates can be added once the normal rate of respiration of the material has been determined. The manometer flask is normally arranged to contain 3 cm³. It is necessary to set up a blank, control manometer, the thermobarometer, which should contain 3 cm³ of distilled water. This thermobarometer registers any temperature or barometric fluctuation and the readings of the other manometers must be adjusted to allow for such changes.

Running technique

When the flasks have been filled, carefully grease their stoppers and spigots with vaseline. Take care not to apply too much; it is best to put on a narrow band over the middle of the stopper, a few twists will spread the vaseline and show whether there is a proper seal. The spigot will require two bands of vaseline, make sure that it is closed and properly sealed. The flasks and spigots are kept firmly in position with rubber bands. Open the main valve at the top of the manometer before placing the flasks in the constant temperature tank (see fig. 4.4).

Before the experiment is started the flasks must be allowed to equilibrate for fifteen minutes in the tank to allow their contents to reach the correct temperature. During this time they should be gently shaken at 120 shakes to the minute. At the end of this time adjust the height of the manometers as required, using the reservoir at the bottom of the manometer (both columns should be high at the start for most respiratory experiments), close the main valve of the manometer, read and record the height of both sides of the manometer. Restart the shaking motor and continue taking readings every fifteen minutes. For all subsequent readings the right-hand column is always adjusted to the same height and only the level in the open, left-hand column need be recorded. The respiratory rate should be steady and clearly established after three or four readings, and then the contents of the side arm, if in use, can be added. Unclamp the whole manometer system, put a finger over the open end of the manometer and carefully tip the flask so that the contents of the side arm flow into the main chamber. Replace the manometer and continue readings as before.

It is possible to continue readings for about two hours before lack of oxygen in the flask begins to affect the results. At the end of the experiment open the top valve to prevent the contents of the manometer sucking back into the flask. Wipe off the vaseline from the spigot and joints using cotton-wool and a little xylol, rinse out in soapy water and leave the flasks and spigots soaking overnight in chromic acid. It is most important to have the flasks properly clean before the start of the next experiment.

A note on the apparatus. Although a complete apparatus, consisting of a thermostatically controlled water-bath together with two sets of seven manometers, can be obtained from biological suppliers, it is an expensive item, and it is possible to construct a useful manometer mostly out of old laboratory equipment. All that is necessary is an electric heating element such as would be used in a kettle, an accurate thermostat, a stirring and shaking motor. Three is the minimum number of manometer units. The shaking and stirring equipment will require some ingenuity in their construction, but the time spent should be well worthwhile.

4. Determination of the respiratory quotient by Warburg manometry (see p. 74)

(see p. 74)

Material. Two lots of 50, 10 mm long root tips of barley. These are obtained by sowing the seed thinly on wide-mesh gauze, which is held above damp blotting paper in a covered dish. Wash them well with distilled water before use and count them out on to moist filter-paper.

Method. As described in exp. 3 (p. 228). The constant-temperature bath should be adjusted to 30°C., and three flasks are filled as follows:

Flask	Purpose	Main chamber	Centre well	Side arm	Total
1	Thermobarometer	3 cm³ water	Nil	Nil	3 cm³
2	O_2 uptake	50 root tips; 2.5 cm³ distilled water	0.5 cm³ strong KOH; filter-paper wick	Nil	3 cm³
3	$\dfrac{CO_2}{O_2}$	50 root tips; 2.5 cm³ distilled water	0.5 cm³ distilled water	Nil	3 cm³

Do not use a mouth pipette for the potassium hydroxide.

The apparatus should be allowed to run for at least an hour. Then corrections for the thermobarometer and the flask constants should be applied and the total oxgen uptake compared with the total carbon dioxide production as in exp. 2 above.

5. Investigation of the effect of addition of an inhibitor on the rate of respiration (see p. 83)

Material. As described in exp. 4 above.

Method. As described in exp. 3 above. Most inhibitors are highly dangerous, so the inhibitor chosen will depend on the local safety regulations. Mercury II chloride, sodium azide, malonic acid or potassium cyanide could be used if regulations permit. The constant-temperature tank should be adjusted to 30°C., and the three flasks are filled as follows:

Flask	Purpose	Main chamber	Centre well	Side arm	Total
1	Thermobarometer	3 cm³ water	Nil	Nil	3 cm³
2	O₂ uptake with inhibitor	50 root tips; 2.0 cm³ phosphate buffer pH 6.5	0.5 cm³ strong KOH; filter-paper wick	0.5 cm³ inhibitor	3 cm³
3	O₂ uptake with inhibitor (10 × diluted)	50 root tips; 2.0 cm³ phosphate buffer pH 6.5	0.5 cm³ strong KOH; filter-paper wick	0.5 cm³ dilute inhibitor	3 cm³

Do not use a mouth pipette for the potassium hydroxide or the inhibitor.

For details of the making up of the phosphate buffer see p. 255. After about an hour, when the respiration rates of the root tips are steady, tip the inhibitor from the side arms. Continue readings for another hour, then, after applying thermobarometer and flask constant corrections, compare the percentage inhibition.

6. Calculation of the flask constants for the Warburg manometer

It is necessary that the flask constants (K) of each flask (together with their own manometer tubes) be calculated if the changes in height of the manometer arm are to be converted from mm to mm³ of gas at N.T.P. Once determined for any particular temperature, they must be carefully recorded and kept for future reference, as they are needed in calculating the results in all manometric experiments.

K can be calculated from the following equation:

$$K = \frac{V_g \frac{273}{T} + V_f \alpha}{P_0}$$

where V_g = volume gas alone in mm³ (not the liquid);
T = absolute temperature;
V_f = volume of liquid in the flask in mm³;
α = solubility of the gas [cm³ at N.T.P. dissolved by 1 cm³ water at 101 kPa (1 atm)];

Take V_f = 3,000;
α O₂   at 25°C = 0.028;
at 30°C. = 0.027;
CO₂ at 25°C. = 0.077;
at 30°C. = 0.067;
P_0 = 10,000;
V_g, the gas volume of the flask + the manometer bore to the zero mark has to be determined, as follows:

Make sure that the flask is dry and clean and weigh it with its spigot to three places. Then fill it with distilled water and insert the spigot. The flask should be filled up to its neck, so that when it is attached to the manometer the liquid rises up into the manometer tube; mark the point to which it rises. Withdraw the flask carefully and reweigh. Calculate the volume of the flask by subtracting the dry weight of the flask from this value. Measure the remaining length of manometer tube to the zero mark and calculate its volume, assuming the bore to have a diameter of 1 mm. Add this volume to that of the flask to obtain V_g.

7. Identification of ethanal (acetaldehyde) as an intermediate in anaerobic respiration (see p. 78)

Material. Freshly washed yeast that has been growing actively. This can be obtained by keeping a culture of baker's yeast in well-aerated water and supplied with glucose and nutrients. Centrifuge this yeast suspension. Reject the supernatant. Resuspend the yeast in distilled water and centrifuge again. Once

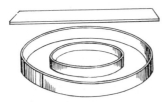

Fig. A.12. The Conway unit.

more reject the supernatant. This process can be repeated until all the nutrients and waste products have been removed.

Method. Place 1 cm³ freshly washed yeast paste in the outer part of a Conway unit (see fig. A.12) add 1 cm³ distilled water and mix. Place 1 cm³ 0.1 per cent 2:4-dinitrophenylhydrazine (*poison, take great care with this substance, do not use a mouth pipette*) in the centre well of the unit. Add 1 cm³ 1 per cent glucose or 1 cm³ 1 per cent 2 oxo-propanoic acid to the outer part and seal the unit immediately, using a little vaseline to ensure an airtight seal. Rock gently to mix the materials in the outer part only and incubate at under 10°C. for about ten hours. Examine for brown crystals of the hydrazone, which are formed if ethanal (acetaldehyde) has been formed by the yeast and has diffused over from the outer part of the unit. If the incubation is carried out at temperatures much higher than 10°C. the 2 oxo-propanoic acid may vaporize and react with the dinitrophenylhydrazine. Careful controls should be set up to check that this does not happen.

8. Identification of some plant acids by simple chemical tests (see p. 79)

Materials. Bryophyllum or *Sedum* spp.
Method. Squash a few leaves in a mortar, filter or centrifuge to clear the extract of chloroplasts and cell-wall detritus. Neutralize the extract carefully with dilute sodium hydroxide and divide into two parts. Test these with the various reagents and note any precipitate (ppt.) formed as indicated in the table on next page.

PLANT ACID	First add a few drops 5% $CaCl_2$	PART 1 Then add an equal quantity of glacial ethanoic acid and boil	PART 1 Then cool and add 95% ethanol	PART 2 First add a few drops 5% lead ethanoate	PART 2 Then add an equal quantity of glacial ethanoic and warm
ethanedioic acid-2-water (oxalic)	White ppt.	White ppt.	White ppt.	White ppt.	White ppt.
2-hydroxybutanedioic acid (malic)	Nil	Nil	Nil	White ppt.	Ppt. dissolves in cold
2-hydroxypropane-1,2,3-tricarboxylic acid (citric)	Nil	Nil	Nil	White ppt.	Ppt. dissolves on warming
2,3-dihydroxbutanedioic acid (tartaric)	Nil	Nil	White ppt.	White ppt.	White ppt. dissolves slightly

9. Identification of plant acids by chromatography (see p. 79)

Material. Leaves of *Bryophyllum* or *Sedum* spp.
Method. Full details are given on p. 210. Squash a few leaves in a mortar, if necessary adding a little 70 per cent ethanol or distilled water. Load plenty of extract on to the chromatogram. The most satisfactory technique is that using butyl methanoate solvent.

10. Investigation of polyphenol oxidase as an example of a terminal oxidase (see pp. 46 and 47)

Material. Fresh potato. Some strains work better than others and younger tubers give better results.
Method.
Preparation of an extract of the enzyme
Cut up half a potato into small chunks, grind in a mortar with 10 cm³ distilled water and filter the extract through glass wool (*care*) at a Büchner funnel. Centrifuge to precipitate the starch grains and cell wall detritus.

Test for polyphenol oxidase activity

The enzyme will cause the conversion of guaiacum (freshly made by dissolving a little guaiacum resin in absolute ethanol to give a pale-brown solution) into guaiacum blue. Add 0.5 cm³ guaiacum to 3 cm³ of the enzyme extract and note the blue colour produced.

Characterization of polyphenol oxidase

(a) EFFECT OF TEMPERATURE ON THE ACTIVITY OF THE ENZYME. Prepare water-baths at 40° and 70°C., place test-tubes containing 3 cm³ enzyme extract in the water-baths and then add 0.5 cm³ guaiacum extract. Note the time taken for the blue colour to appear and compare with the time taken at room temperature. How does temperature affect the activity of the enzyme?

(b) EFFECT OF pH ON THE ACTIVITY OF THE ENZYME. Prepare phosphate buffer solutions (see p. 255) of pH 3.0, 5.5, 7.0, 8.0. Place 2 cm³ of each buffer in separate test-tubes and add 1 cm³ enzyme extract and 0.5 cm³ guaiacum to each. Note the colours produced over a period of time. At what pH does the enzyme operate most efficiently?

11. Investigation of the dehydrogenase activity of etiolated pea shoots (see pp. 83)

Material. A washed suspension of yeast (see p. 231); pea seedlings grown in the dark for some days.
Method.
Preparation of an extract of the enzyme from pea seedlings
Squash up a few etiolated shoots in a mortar with 10 cm³ ice-cold sucrose buffer* or phosphate buffer pH 6.5.* Centrifuge at medium speed for five minutes to

clear the extract and treat the supernatant as follows. The washed yeast suspension may be used direct.

(a) TEST FOR DEHYDROGENASE ACTIVITY. Dehydrogenase enzymes cause bleaching of redox dyes such as dilute alkaline methylene blue or better, very dilute 2:6-dichlorophenolindophenol (DCPIP)† at the more satisfactory pH 6.5. Such bleaching occurs more quickly if no oxygen is present; accordingly, it is best to operate the following tests in small-sized test-tubes, which give less surface area for oxidation, or else in special Thunberg tubes (see fig. A.13). The air can be removed from such tubes using a vacuum pump.

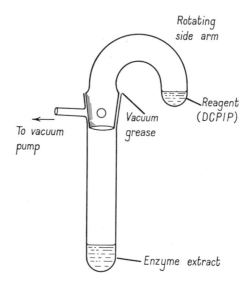

Fig. A.13. The Thunberg tube. A device for estimating the activity of dehydrogenase enzymes.

Take 1 cm³ of the enzyme extract or suspension and place this in the main part of the Thunberg tube and add 0.5 cm³ DCPIP to the side arm. Adjust the side arm so that the tube can be evacuated and connect it up to a vacuum pump. After a few minutes close the tap by rotating the side arm and disconnect the vacuum pump. Tip the DCPIP dye from the side arm into the main part of the tube and take the time for the dye to become reduced and bleached.

(b) EFFECT OF TEMPERATURE ON THE ACTIVITY OF THE ENZYME. Repeat the experiment described in the above section at a series of temperatures, e.g. 30°, 40°, 50°, 60°C., using a water-bath to obtain the desired temperature and allowing the solutions in the Thunberg tube a few minutes to reach the correct temperature before tipping the side arm. Record the time taken for the bleaching to occur in each case. Plot 1/time against temperature on a graph.

---

† Note: this is also a test for ascorbic acid (vitamin C), which may interfere with the reaction.

12. Spectroscopic examination of the cytochromes (see p. 85)

Material. Actively growing baker's yeast.

Method. Centrifuge some freshly grown yeast and resuspend the precipitated paste in cool, freshly boiled water. Add a little 10 per cent sodium hydrosulphite, a mild reducing agent to help intensify the bands and recentrifuge. Spread the paste on a slide so as to form a layer at least 2mm thick, cover with another slide and illuminate with a high-intensity microscope lamp. View through a low-dispersion hand spectrometer and note the incidence of absorption bands. These may be visible as follows:

Part of the spectrum	Wavelength in nm	Strength	Type of cytochrome	Absorption band
Red	603	Weak band	Cytochrome *a*	α
Yellow	563	Strong band	Cytochrome *b*	α
Yellow–green	550	Very strong band	Cytochrome *c*	α
Green–blue	c.525	Weak band	Cytochrome *a, b, c*	β

APPENDIX TO CHAPTER 5

1. The use of water cultures for investigating the plant's requirement for various minerals
2. The culture of fungi on agar media for investigating their mineral requirement
3. Determination of the effect of anaerobic conditions and low temperatures on the absorption of bromide ions by excised barley roots
4. Determination of the effect of anaerobic conditions and low temperature on the absorption of ammonium ions by carrot discs, using the Conway method for estimating ammonia
5. Investigation of the salt respiration effect using the Warburg manometer
6. Investigation of the rate of uptake of dissolved mineral by the use of ^{32}P radioactive tracer
7. Investigation of the path of movement of dissolved mineral in the stem by the use of ^{32}P radioactive tracer
8. Investigation of the areas of accumulation of phosphate by autoradiographic analysis of ^{32}P

1. The use of water cultures for investigating the plant's requirement for various minerals (see p. 92)

Material. Barley seed.

Method.

(i) *Preparation of young seedlings*

Take an old cork mat and drill a series of small holes of about 0.5 cm in diameter through it. Paint wax over the whole mat and float it in a trough of water. Place a seed over each hole and leave them to germinate. After about ten days the leaves should be quite well developed; transfer the seedlings to the culture solution bottles

(ii) *Preparation of the culture bottles*
Half-pint milk bottles make useful containers. Wash seven of them well with boiling water, followed by distilled water, and then fit them with carefully cleaned corks which have been drilled with 1 cm-diameter holes. If an air pump is available it is a great advantage to provide the bottles with aeration tubes, as growth rates are more than doubled if the roots are kept well aerated.

(iii) *Preparation of the culture solutions*
The complete medium is made up as follows:

(From James: *An Introduction to Plant Physiology*, by permission of the Clarendon Press, Oxford.)

$CaSO_4.2H_2O$	0.25 g
$Ca(H_2PO_4)_2.H_2O$	0.25 g
$MgSO_4.7H_2O$	0.25 g
NaCl	0.08 g
KNO_3	0.70 g
$FeCl_3.6H_2O$	0.005 g

Make up to 1 dm^3.
For cultures lacking various elements substitute as follows:

Potassium:	replace	KNO_3	by	0.59 g $NaNO_3$
Calcium:	replace	$\begin{cases} CaSO_4.2H_2O \\ Ca(H_2PO_4)_2.H_2O \end{cases}$	by by	0.20 g K_2SO_4 0.71 g $Na_2HPO_4.12H_2O$
Iron:	omit	$FeCl_3.6H_2O$		
Nitrogen:	replace	KNO_3	by	0.52 g KCl
Phosphorus:	replace	$Ca(H_2PO_4)_2.H_2O$	by	0.16 g $Ca(NO_3)_2$
Sulphur:	replace	$\begin{cases} CaSO_4.2H_2O \\ MgSO_4.7H_2O \end{cases}$	by by	0.16 g $CaCl_2$ 0.21 g $MgCl_2$

Fill the bottles with the required nutrient, cover them with black paper to reduce algal growth and label them. Place a barley seedling in each so that the seed is just below the top of the cork and keep it in place with a pad of dry cotton-wool. Place the bottles in a well-lit place, turn on the aerating system, if it is available, and leave for some weeks. The cotton-wool will probably become damp, so it should be renewed as often as possible to reduce the risk of fungal infection. The culture solutions should be renewed every fortnight, and loss due to evaporation and transpiration should be made good by topping-up with distilled water.

2. The culture of fungi on agar media for investigating their mineral requirement (see p. 92)

Material. Mucor or any actively growing species.
Method.
Preparation of general nutrient agar
It is first necessary to grow the fungus so as to obtain it in an actively growing, uniform and pure state. A most useful medium for this purpose is known as potato agar and is made up as follows:

Take 200 g of clean peeled potatoes and chop them up as finely as possible into 200 cm^3 of water. Boil for about half an hour and then allow to settle and cool.

Filter through muslin and make the solution up to 400 cm^3 in a large flask, add 10 g of agar and place in an autoclave at 200 kPa above atmospheric pressure (30 lb. in^{-2}) for ten minutes. The medium is then sterile and ready for pouring on to Petri dishes that have been sterilized previously. Once the plates have cooled and the agar set, they can be inoculated by means of a sterile needle touched on to the surface of the stock fungus culture. When fungi are first isolated from wild cultures they grow much better on the above medium, but for the investigation of their mineral requirements it is necessary to make up a special medium:

Preparation of special media for investigating the requirement of the fungus for various minerals

Add 10 g of agar and 20 g of glucose to 400 cm^3 of distilled water and heat until they are dissolved. The following nutrients are required for a complete medium of 400 cm^3. Different nutrient media can be made up by omitting any one mineral and substituting another as in exp. 1 above.

$(NH_4)_2HPO_4$	0.5 g
KNO_3	0.4 g
$CaCl_2$	0.2 g
$MgSO_4.7H_2O$	0.2 g
$FeCl_3.6H_2O$	0.002 g
Biotin (vitamin B1 or thiamin)	0.5 g

Add the desired minerals and sterilize the media as above. Pour the sterile agar on to sterile Petri dishes and inoculate with fresh and actively growing fungus. Incubate at 25 °C. Growth should start in about two days, and measurements of the size of the cultures can be made over a period of time; for this, measurement of the diameter of the colony is sufficient, in most cases.

Note. For full details of a wide variety of media used for culturing microorganisms, see *The Oxoid Manual of Culture Media*, 2nd edn. (1962), issued by the Oxoid Division, Oxo Ltd., Southwark Bridge Road, London S.E.1.

3. Determination of the effect of anaerobic conditions and low temperatures on the absorption of bromide ions by excised barley roots (see p. 100)

Material. Actively growing barley seedlings (see exp. 4 on p. 229).
Alternatively, washed carrot discs may be used (see exp. 4 below).
Method. Cut off 300 tips of barley root 10 mm long into a beaker of distilled water. Wash well in distilled water at a Büchner funnel and count on to moist filter-paper so that they can be transferred to the solutions. One hundred cm^3 of 0.005 M potassium bromide are pipetted into each of three large conical flasks. Flasks one and two are aerated by a motor, and flask two is placed in a trough packed with ice and maintained at about 3 °C. Flask three is bubbled through with nitrogen from a cylinder; the solution in this flask should previously be boiled under reduced pressure to remove dissolved oxygen. It must be brought back to atmospheric pressure by allowing nitrogen to bubble into the solution. One hundred root tips are then placed in each flask.

While the experiment is running the strength of the 0.005 M KBr can be checked and the titration technique perfected. Pipette 25 cm³ of the KBr into a stoppered flask, add 2.5 cm³ ethanoic acid so that the acid is 0.1 M before titration and finally add 6 drops of 2 per cent aqueous eosin as an indicator. Titrate against 0.02 M $AgNO_3$, shaking the solution vigorously to facilitate the absorption of the eosin upon the precipitated AgBr. The end point is reached when the AgBr precipitate turns magenta. Calculate the g/dm³ of bromide in the stock solution. After about three hours take the temperature of the experimental flasks, decant off the liquid and adjust the volume of each flask to allow for any evaporation that may have taken place, and determine how much bromide has been absorbed by the root segments in each case. Lay out your results in table form, expressing the uptake of bromide under nitrogen and low-temperature conditions as a percentage of the absorption in air at room temperature.

(After Brierley (1958). *School Science Review*, No. 138, p. 254.)

4. Determination of the effect of anaerobic conditions and low temperature on the absorption of ammonium ions by carrot discs, using the Conway method for estimating ammonia (see p. 100)

Material. Thirty uniform carrot discs, washed in running water for about twelve hours.

Method. Set up three flasks as in exp. 3, but fill each with 100 cm³ 0.02 M ammonium phosphate $(NH_4)_2HPO_4$. Place ten weighed discs in each flask, cork up and adjust the bubbling gas. After about six hours the discs may be filtered off and the concentration of ammonium estimated. Meanwhile the 0.02 M ammonium phosphate should be checked and the estimation technique perfected.

The Conway method for estimating ammonia

Pipette 2 cm³ of the ammonium solution into the outer chamber of the Conway unit (see fig. A.12). Pipette 1.5 cm³ borate buffer (see p. 253) into the centre well, grease the rim of the unit and add 1 cm³ strong sodium hydroxide to the outer chamber using a pipette fitted with a rubber bulb. Seal the unit, rock it gently (do not mix the inner and outer liquids) and incubate at 40°C. for about two hours. Then remove the cover and add 0.02 M HCl from a burette to the centre well until it regains its original colour (i.e. it turns back, from blue to pink). Calculate the efficiency of 'recovery' of ammonia and make the necessary allowance when titrating the unknown samples. Record your results in table form, expressing the uptake of ammonium under nitrogen and low-temperature conditions as a percentage of the absorption in air at room temperature. Allow for any small weight differences in the carrot discs.

5. Investigation of the salt respiration effect using the Warburg manometer (see p. 100)

Material. A culture of the yeast-like fungus, *Torilopsis utilis*, or similar species, which has been growing on nitrogen-deficient medium.

Method. As described in exp. 3 (p. 228) above. The constant-temperature tank should be adjusted to 25°C. and the three flasks are filled as follows:

Flask	Purpose	Main chamber	Centre well	Side arm	Total
1	Thermobarometer	3 cm³ water	Nil	Nil	3 cm³
2	O_2 uptake: NH_3 sampled at end	2.2 cm³ yeast suspension	0.3 cm³ strong KOH; filter-paper wick	0.5 cm³ 0.03M $(NH_4)_2HPO_4$	3 cm³
3	O_2 uptake; NH_2 sampled at end	2.2 cm³ yeast suspension	0.3 cm³ strong KOH; filter-paper wick	0.5 cm³ 0.03M Na_2HPO_4	3 cm³

Do not use a mouth pipette for the potassium hydroxide

After about an hour, when the respiration rates of the yeast are steady, tip on the phosphate from the side arms. Continue readings for another hour, then dismantle the flasks. After applying thermobarometer and flask constant corrections, compare the percentage stimulation on addition of sodium and ammonium phosphate.

The ammonium content of the yeast and the external solution are then determined for each flask. In flask 2 (which should have shown a rise in the respiration rate on the addition of ammonium phosphate) the fungus should have taken up a considerable quantity of ammonia, thus leaving the external solution less concentrated. Flask 3, which should not have shown much respiratory response on the addition of the sodium phosphate, acts as a control, and the quantity of ammonia normally present in the yeast can be determined so that the actual take-up in flask 2 can be found. The ammonia can be determined by the Conway method (see exp. 4 above).

6. Investigation of the rate of uptake of dissolved mineral by the use of ^{32}P radioactive tracer (see p. 103)

Material. Two young rooted plants of tomato, *Zebrina* (*Tradescantia*), balsam (*Impatiens sultani*), or of a woody perennial, such as willow or *Skimmia*. If possible these cuttings should be rooted in water or nutrient solution, as it is difficult to avoid damage if they are dug up from soil. If they have been grown in soil, transference to nutrient medium a few weeks before the experiment will be satisfactory. Aeration considerably helps growth under these conditions.

Method. First cover the bench top with thick, waterproof paper. All apparatus should stand on enamel trays. Make up the test solution by preparing 100 cm³ 0.07 M phosphate buffer (see p. 255) and add sufficient radioactive phosphate to give the solution a strength of between 3 and 6 micro-curies (µCi). Take care when pipetting out the solution of ^{32}P; use rubber gloves and a special pipette with a bulb or piston system for sucking up the liquid, *under no circumstances use an ordinary mouth pipette.* Cut a thin piece of lead to cover the culture jar (see fig. 5.12), leaving a slit for the plant stem. Wedge the plant in carefully with a small piece of cotton-wool and fill the jar with the tracer solution. Prepare a similar control specimen, but omit the ^{32}P; leave both plants in a well-illuminated and aerated place. If the plants' roots are small, the tracer may be placed in a small tube, such as a specimen tube, inside the larger jar. The specimen tube may be kept in place by means of Plasticine.

Using a ratemeter, estimate the counts per minute at a given position, say 2 cm from the main apex; the control is useful for comparing the background counts of the two specimens at the start. Record at frequent intervals until the counts per minute reach several hundred. This will take about one day for the annuals, usually longer for woody plants. Plot the counts per minute against time on a graph, and thus obtain an idea of the rate of absorption of the phosphate. The experiment could be repeated under different external conditions to find the effect of environmental factors on the rate of uptake.

For full details of regulations and precautions see p. 222. ^{32}P, although a fairly strong β-emitter (1.71 MeV), has most of its radiation absorbed by glassware. Its short half-life (14.2 days) makes waste disposal simple as solutions may be stored until their activity is negligible.

7. Investigation of the path of movement of dissolved mineral in the stem by the use of ^{32}P radioactive tracer (see p. 102)

Material. Two rooted cuttings of a woody plant prepared as in the above experiment.

Method. Make up the nutrient solution as described above, but so as to produce 200 cm^3 solution with an activity of 6 μCi. Place 100 cm^3 of the tracer solution in each of two small jars and insert a rooted plant in one jar and a similar but ringed plant in the other, enclosing each with lead as in exp. 6. When ringing the stem take a scalpel and cut through the bark and phloem so as to make two rings 1 cm apart. Remove the bark in between and scrape off any phloem and cambium (the slimy tissues) remaining. Estimate the counts per minute as in exp. 6. Has ringing made any difference to the accumulation of phosphate in the upper parts of the plant? Record the uptake against time on a graph for the two plants.

8. Investigation of the areas of accumulation of phosphate by autoradiographic analysis of ^{32}P (see p. 104)

Materials. As produced in exp. 6 and 7 above.

Method. Remove the specimen from the culture jar, rinse off the nutrient solution thoroughly and dry it with tissue (use rubber gloves and dispose of the tissues in the bin allocated for radioactive waste). Alternatively, cut off the aerial parts and analyse these alone. Make an autoradiograph of the specimens as described on p. 223. Exposure time will be about 36 hours, with Kodirex and material giving a count of about 1000 per minute 2 cm from the material. What conclusions can you draw about the distribution of ^{32}P in the stem, apex and leaves of the specimen?

APPENDIX TO CHAPTER 6

1. Squash preparations of chromosomes
2. Simple tests for proteins

3. Testing for the presence of DNA and RNA
4. Chromatographic identification of amino-acids and proteins
5. Identification of simple sugars by chromatography
6. Investigation of the hydrolysis of urea by urease
7. Polarimetric study of the inversion of sucrose
8. Extraction, purification and properties of starch phosphorylase
9. Identification of structural carbohydrates
10. Identification of phenolic substances
11. Investigation of the cyanogenic properties of plant leaves
12. Identification of lipids

1. Squash preparations of chromosomes (see p. 115)

Materials. Onion (*Allium cepa*) is very useful for root-tip preparations. Anthers of *Tradescantia, Fritillaria, Allium* and *Paeonia* spp. are suitable for meiosis. (Mitosis is also frequently observed in the developing anther.)

Method. This method normally uses *no* fixation. Alternatively, 24 hours fixation in Carnoy le Brun (1 part glacial ethanoic (acetic) acid, 1 part trichloromethane, 1 part absolute ethanol) can be used and gives good results, especially with pollen-grain meiosis material.

Cut the anther or root tip from the living plant and warm it in aceto-lacmoid+HCl* stain for five minutes. Take care to avoid getting any chromosomal stain on your skin. A watch-glass and spirit lamp are useful for this purpose.

Cut off a small part of the end of the root tip or part of the anther; macerate it with a needle and mount under a round No. 1 cover-slip in aceto-orcein* stain. Tap with the wooden end of the needle, using a blotting-paper pad over the cover-slip to spread the material. Care must be taken not to move the cover-slip. Warm gently two or three times during the tapping to help spread the material and lessen cytoplasmic staining.

When examining the preparation correct lighting is important and green filtering is most beneficial. The best squashes should be examined with an oil-immersion lens.

Permanent preparations

These can be made by first smearing egg albumen over the cover-slip and warming it. The material will then stick to the cover-slip, and this may be removed by floating it off in 45 per cent ethanoic acid. Dehydrate quickly and mount in Euparal.

Note : If fresh material is used most of the staining is done by the aceto-orcein; if fixed material is used the lacmoid is the main staining agent. In the first case the HCl in the lacmoid is important in aiding the penetration of the stain by macerating the tissues. Lack of fixation sometimes results in rather bubbly chromosomes, which occasionally appear double; care is required in interpreting such preparations.

2. Simple tests for proteins (see p. 241)

Materials. Protoplasm is made-up largely of proteins and amino-acids, so that considerable quantities are found in any living cell or tissue; they are particularly concentrated in rapidly growing meristematic areas, especially the stem and root apices.

Tests

(i) MILLON'S TEST (Cole's version).* Add a little Millon's 'A'* (*care: this contains mercury*) to the protein, boil, allow to cool; a pale-yellow colour indicates a considerable quantity of protein. Add a drop of Millon's 'B',* a red colour, produced on warming, indicates the presence of the amino-acid *tyrosine*, which is common both as a constituent of proteins and also as a free amino-acid.

(ii) XANTHOPROTEIC TEST. Place about 2 cm³ of the protein solution in a test-tube and add a few drops of concentrated nitric acid. A cloudy precipitate is formed which turns yellow on boiling. After cooling add ammonium hydroxide until the solution is alkaline, when it will turn orange. This is a general test for proteins.

(iii) BIURET TEST. Warm the tissue to be tested in a little distilled water. To this add 2 cm³ of 40 per cent sodium hydroxide and one drop of 1 per cent copper sulphate. A pale lilac colour, produced on warming, indicates proteins.

(iv) IODINE DISSOLVED IN POTASSIUM IODIDE. This gives a golden-brown colour with proteins, but also with several other substances; it is, however, particularly useful in helping to show up the cytoplasm and nucleus.

3. Testing for the presence of DNA and RNA (see p. 183)

Material. Root tips of onion or hyacinth.

Method. Remove the growing root tips from the plant and fix them overnight in absolute alcohol. Using a razor cut thin sections of the terminal 5 mm of the tip and stain these in methyl-green-pyronin aqueous stain* for 30 minutes. Rinse in distilled water and mount in water under a coverslip. Examine under medium and high power; DNA will be stained green and RNA red.

(After Nuffield Advanced Science (Biology) (1970). *The Developing Organism*, p. 52. Harmondsworth, Penguin.)

4. Chromatographic identification of amino-acids and proteins
(see p. 123)

Materials. Young and actively metabolizing tissues such as seedlings and root tips.

Method. Grind up the tissue in a mortar with a little 80 per cent ethanol and spot the extract on to the chromatogram. The chromatography technique is as described p. 217, the best solvent being phenol.

Amino-acids will appear without special treatment; proteins themselves will not show up on the chromatogram, but their amino-acids can be released by heating the original extract to 100°C. with 6 M hydrochloric acid in a sealed tube for a few minutes. This operation should be carried out with great care and only by experienced staff.

5. Identification of simple sugars by chromatography (see p. 210)

Material. Fresh onion or spinach leaves.

Method. Place two batches of leaves in water and leave them in the dark for twelve and twenty-four hours; in this way they will become progressively starved. Place a third batch in the light to allow for the accumulation of higher carbohydrates.

After light and dark treatments make extracts of the material by squashing similar quantities of the leaves in a mortar with a little 50 per cent ethanol. Apply the extracts to filter-paper strips, sheets or thin-layer plates. The most satisfactory solvent is probably phenol. Full details are given on p. 212. In addition to identifying the spots, make an estimate of the relative amount of each substance found.

6. Investigation of the hydrolysis of urea by urease

Material. Standard tablets (B.D.H.) of urease.

Method. Urea is converted by the enzyme to ammonium carbonate which can be titrated with sulphuric acid to obtain a quantitative estimate of the activity of the enzyme preparation in a variety of conditions of temperature and pH.

Place 10 cm^3 of 1 per cent urea in water in a test tube. Crush up one standard tablet of urease in 3 cm^3 of water (or phosphate buffers at different pH*) and place this in a second test tube. Put both tubes in a water bath at the required temperature (say 25°C.). Leave for a few minutes for the tubes to adjust to the temperature and then pour the urea solution into the urease, note the time and leave for 45 minutes. Decant the fluid off rapidly through a small pad of cotton-wool and pipette off 10 cm^3 of the clear solution and place in a conical flask. Titrate against 0.05 M sulphuric acid using methyl orange as indicator. 1 cm^3 of the sulphuric acid is equivalent to 3 mg urea.

(After James, W. O. (1973). *An Introduction to Plant Physiology.* 7th edn. Clarendon Press, Oxford.)

7. Polarimetric study of the inversion of sucrose (see pp. 136 and 137)

The Polarimeter

The polarimeter (fig. A.14) is an instrument designed to determine the degree

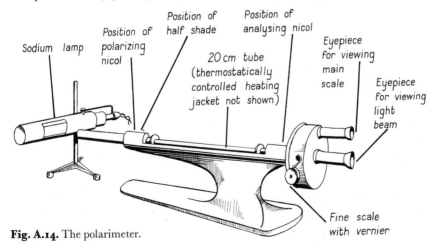

Fig. A.14. The polarimeter.

certain substances are capable of rotating the plane of polarized light. It consists essentially of a source of monochromatic light, usually a sodium lamp or flame, secondly, a polarizing Nicol prism, thirdly, a tube, usually 20 cm (2 dm) long which contains the solution being investigated, and finally, an analysing prism. The instrument is provided with a *half-shade* device which makes analysis of the plane of rotation easier, the reading being taken when both sides of the field of view are equally illuminated. The plane of rotation is recorded on a vernier scale. A good instrument is also provided with a thermostatically controlled water jacket around the solution tube so that its temperature can be carefully regulated.

Setting up the polarimeter

First check the apparatus and see that the scale is on the zero mark; make adjustments if necessary. Make sure that the tube is clean and then fill it with the solution being examined, taking care to avoid bubbles. Do not screw down the end pieces of the tube too tight. Before placing the tube in the polarimeter check that it does not leak (and check again, at intervals, during the course of the experiment).

To determine the rate of inversion of sucrose at 20°C. using dilute acid

Prepare the following solution:
1. 20 g sucrose dissolved in 100 cm³ distilled water.
2. 50 g M hydrochloric acid.
Place these in an incubator at 20°C. for about half an hour.

Set up the heat jacket and pump system so as to provide a temperature of 20°C., and check the optical rotation of the 20 per cent sucrose. Knowing the specific rotation of sucrose at 20°C. to be $+66.5°$, then the expected rotation can be calculated from the formula:

$$\left[\alpha\right]_D^{20} = \frac{100.\alpha}{l.c}$$

where $\left[\alpha\right]_D^{20}$ = Specific dextro rotation at 20°C.;

α = observed rotation; l = length of tube in dm; c = g sugar in 100 cm³ water. In this case the rotation should therefore be $+26.6°$.

After checking the rotation of sucrose, add 1 cm³ acid to 20 cm³ sucrose solution and record the rotation over a period of half an hour.

To determine the rate of the inversion of sucrose at 20°C. using the enzyme invertase

Dissolve 20 g of sucrose in 100 cm³ 0.1 M NaH_2PO_4. This is incubated at 20°C. for half an hour and its rotation checked as above; 1 cm³ of invertase concentrate (or as directed) is added to 20 cm³ of the sucrose solution and the rotation recorded over a period of about half an hour. In this time the rotation should be reduced to zero.

Compare the graphs showing the rate of inversion using mineral acid and enzyme. By varying the temperature of the outside jacket (and temperature at which the materials are incubated), the temperature optimum for the inversion process may be found.

8. Extraction, purification and properties of starch phosphorylase (see p. 27 and 138)

Material. Most varieties of potato; peas (round varieties contain more starch phosphorylase than wrinkled varieties).

Method. Chop up and grind a small potato in a mortar together with a little ice-cold phosphate buffer pH 7.0 (see p. 255) so as to produce about 10 cm³ extract. Decant carefully and centrifuge at medium speed for about 5 minutes to clear the extract of starch grains. Check a drop of the supernatant for the presence of starch, using iodine. If necessary, centrifuge again. Pour off the supernatant and keep it cool until you are ready to carry out the next stage.

Make up 2 cm³ of a 1 per cent solution of glucose-1-phosphate in distilled water. This must be freshly prepared.

Prepare two test-tubes with a 1 cm³ glucose-1-phosphate in each. Add 1 cm³ of the enzyme extract to the first tube and 1 cm³ boiled extract to the second, shake the tubes. Test for the presence of starch immediately and at four-minute intervals by removing a drop of the liquid with a pipette and placing it on a white cavity tile together with a drop of iodine solution. Check that the original enzyme extract is starch free. How long does the starch take to be formed? Is the enzyme inactivated by boiling?

9. Identification of structural carbohydrates (see p. 139)

(a) *CELLULOSE*

Materials. The cell walls of most plants, particularly the collenchymatous or thickened walls found towards the outside of the cortex of some stems (e.g. *Helianthus*).

Test. This is a structural carbohydrate composed of complex chains of hexose units; it does not react with iodine unless first treated with concentrated sulphuric acid, when a blue-violet colour will be obtained (the Amyloid reaction). A better method is to treat a thin section of the material with Schultze's reagent,* which gives a blue colour with cellulose, but, like the Amyloid reaction, also stains any starch that may be present.

(b) *HEMICELLULOSES*

Materials. The cell walls of many seeds, e.g. the garden lupin and nasturtium (*Tropaeolum*).

Test. Superficially they resemble cellulose and they form thickenings on the inside of the cell wall, but they are easily hydrolysed by dilute acid and can be distinguished from cellulose in this way.

10. Identification of phenolic substances (see p. 142)

(a) *LIGNIN*

Materials. Woody tissues, xylem, fibres and sclerenchyma.

Test. It is stained bright magenta-red by phloroglucinol with HCl* and yellow by aniline sulphate.*

(b) *ANTHOCYANIDIN AND FLAVONOID PIGMENTS* (see p. 145)

Materials. Coloured flowers; leaves in autumnal colours.

A. *Chemical identification*

Grind a few petals or leaves in 80 per cent ethanol to obtain a general extract. Centrifuge or filter to remove debris and place in a separating funnel with an equal volume of petroleum ether (B.P. 100–120°C.). Shake and then separate. Test the layers as follows:

(*a*) UPPER LAYER (petrol ether). If this is yellow, *carotenoids are present.* Add an equal volume of 95 per cent ethanol, shake and examine the distribution of colour. *Xanthophylls* are more soluble in the ethanol layer. *Carotenes* remain in the ether layer.

(*b*) LOWER LAYER (ethanol). If *coloured* this may contain both *anthocyanins* and *flavonoids*, if *colourless* no *anthocyanins* but possibly *flavonoids*.

These can be separated by dividing the solution into five test-tubes, having 2 cm^3 extract in each. Treat each as follows:

1. Add 4 drops 1 per cent dilute HCl to make the solution acid.
2. Leave (neutral).
3. Add 4 drops 1 per cent Na$_2$CO$_3$ to make the solution alkaline.
4. Add 4 drops 1 per cent NaOH to make the solution alkaline.
5. Make the solution alkaline with a few drops of 1 per cent Na$_2$CO$_3$, and then add 4 drops dilute FeCl$_3$.

The following table summarizes the colours that will be obtained with pure anthocyanins and flavonoids and with mixtures of the two:

	1 ACID	2 NEUTRAL	3 Na$_2$CO$_3$	4 NaOH	5 FeCl$_3$
AC	Red to mauve	Blue to mauve to pink	Blue	Blue	Slate–blue to purple
F	Yellow or colourless	Yellow or colourless	Yellow	Yellow	Olive–yellow
AC+F	Red	Pink	Green	Olive–brown–yellow	Olive–brown

There are no simple tests to distinguish the type of anthocyanidins present, but the following are frequently found, the colours being when the solution is neutral:

1.	*Pelargonidin*	Scarlet-red	5.	*Cyanidin*	Purple
2.	*Peonidin*	Crimson	6.	*Hirsutidin*	Blue
3.	*Malvidin*	Mauve	7.	*Delphinidin*	Blue
4.	*Petunidin*	Purple			

B. *Chromatographic identification*

Full details of the filter-paper chromatography technique are given above on p. 210. An extract of the pigments is made in 80 per cent ethanol and the filter-paper spotted in the usual way; do not leave the extract, as it will fade. No developing spray is necessary.

Unfortunately it is extremely difficult to obtain reproducible results, and it is best to run the unknown extract against a known one such as can be obtained from the scarlet geranium (*Pelargonium*). In this case it is best to use a larger jar (see fig. A.6, p. 000), so that several different spots can be compared at the same time. The ratio of the R_f of the pigment to that of *Pelargonidin* should be a constant.

Intensification of the spots is possible by use of strong ammonia vapour. This will, of course, change the colour of the spot, but will, in addition, show up the presence of yellow Flavonoids, which have R_f values between 0.9 and 1.0.

Investigation of the cyanogenic properties of plant leaves (see p. 147)

Material. White clover (*Trifolium repens*), Birdsfoot-trefoil (*Lotus corniculatus*), Cherry-laurel (*Prunus laurocerasus*).

Method. Cut off about 3 cm² leaves (about three complete leaves, i.e. nine leaflets in all of *Trifolium*) and place these in a specimen tube together with one drop of toluene (methylbenzene) (*caution: this substance is toxic, do not get it on your hands or inhale the vapour*). Grind the leaves gently with a *clean* glass rod (it must not have been in contact with other plants). Moisten a piece of sodium picrate paper* (*caution: these papers are poisonous, handle them with tweezers*) with distilled water and cork the specimen tube so that the paper hangs over the leaves without touching them (see fig. A.15). Incubate the tubes for one hour at 40°C. (or overnight in a

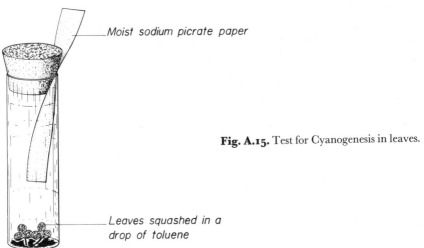

Moist sodium picrate paper

Fig. A.15. Test for Cyanogenesis in leaves.

Leaves squashed in a drop of toluene

warm place) and note any change in colour in the paper. If the yellow paper has changed to an orange colour then cyanide has been formed in the leaves. Compare leaves from different plants of the same species

12. Identification of lipids (see p. 142)

Unlike the carbohydrates, the lipids are insoluble in water but soluble in ether or chloroform.

(a) OILS AND FATS

These are the triglycerides of fatty acids and are common food reserve substances.

Materials. Leaves of cocksfoot (*Dactylis glomerata*) and water mint (*Mentha aquatica*). Fruits of the olive (*Olea europaea*) and in the endosperm of many seeds such as the coconut (*Cocos nucifera*), sunflower (*Helianthus annuus*) and linseed or flax (*Linum usitatissimum*).

Test. Fat globules are stained by various Sudan stains.

(i) *Sudan II or IV.* (0.5 per cent in 70 per cent ethanol.) Wash excess stain from the material with 90 per cent ethanol. Fat globules are stained red.

(ii) *Sudan Black.* (1 per cent in 95 per cent ethanol or in propan-1,3-diol, warm on a water-bath to dissolve the solid; filter before use.) This stain is useful for bulky tissues. Remove excess stain by washing the material with 70 per cent ethanol.

(iii) *Sudan Blue.* (0.5 per cent in ethanol; warm on a water-bath to dissolve the solid; filter before use.) This stain is useful for microscopic material. Remove excess stain by washing the material with distilled water.

(b) WAXES

These differ from the fats and oils in that the glycerol is replaced by monohydric or occasionally dihydric alcohols.

Materials. Leaves of most plants, particularly conifers, stems of cacti (*Cereus* spp.).

Test. These are not found in globules, but cover the epidermis of many plants. They are also stained by the Sudan stains.

(c) CUTIN AND SUBERIN

These are wax-like substances consisting of a mixture of various waxy condensation products of various fatty acids.

Materials. Cutin is found in the cuticular layer of many plants such as the *Rhododendron* and cherry-laurel. Suberin is the waterproofing substance in cork, and is common in the outer layers of old stems and roots.

Test. These substances also react with the Sudan stains.

APPENDIX TO CHAPTER 7

1. Investigation of the effect of auxins on the growth of cress roots and stems
2. Investigation of the effect of gibberellic acid on the elongation of stems
3. Investigation of the effect of gibberellic acid on the activity of the enzyme amylase.

1. Investigation of the effect of auxins on the growth of cress roots and stems (see p. 160)

Material. Cress seed.

Method. Sterilize twelve Petri dishes and surface sterilize 120 seeds by shaking them for three minutes in a 1 per cent solution of sodium hypochlorite. Wash them thoroughly in distilled water. Place a filter-paper in each Petri dish and moisten it with 5 cm³ distilled water. Count ten seeds on to each filter-paper and place the Petri dishes in a dark incubator at 30°C. for forty-eight hours.

Prepare stock solutions of indolylacetic acid (IAA),* and indolylbutyric acid (IBA). Dilute the stock solutions so as to provide the following working solutions:

(i) Control (distilled water)
(ii) 0.01 p.p.m.
(iii) 0.1 p.p.m.
(iv) 1.0 p.p.m.
(v) 10 p.p.m.
(vi) 100 p.p.m.

After forty-eight hours remove the dishes and select the six most uniform. healthy seedlings in each dish, rejecting the less healthy; make careful, *separate*, measurements of the root and stem length of each seedling, the root length should be about 10 mm.

Add 7 cm³ of the required auxin solution to each dish so as to provide a range of seedlings with different auxin concentrations. Replace in the incubator and after two days re-measure the stem and root lengths. Calculate the average increase in length for each auxin concentration and plot your results on a graph. Compare the stimulating and inhibiting effects of the two auxins.

2. Investigation of the effect of gibberellin on the elongation of stems
(see p. 166)

Material. Young plants of a variety of species e.g. *Bryophyllum*, cabbage, the dwarf variety of the garden pea *Meteor* is particularly useful.

Method. Make up a solution of GA by diluting the stock solution* 20 × to give a solution of 1 part in 20000. Apply one drop to the apex of the plant and continue application every four or five days for about three weeks. A control plant should be treated similarly but with water. Record the growth of the treated and untreated plants and also the number of leaves on each plant. Treatment of tall growing strains of pea e.g. Pilot, has little effect.

(Modified after James, W. O., *An Introduction to Plant Physiology*, 7th edn. Exp. 33, p. 165. Clarendon Press, Oxford)

3. Investigation of the effect of gibberellin on the activity of the enzyme amylase (see p. 166)

Material. Barley seed.

Method. Prepare two sterile Petri dishes with a 3 mm layer of starch agar (0.5 per cent starch in 1 per cent agar) and allow to set. Prepare two more dishes with a similar medium to which has been added 1 cm³ GA stock solution so as to give a final concentration of 10 parts per million. The GA should be stirred in while the agar is still liquid.

Strip the husks from twenty barley grains, soak for one hour and then cut them across into two halves, so that one half contains the embryo. Surface sterilize the pieces by dipping them in a one per cent solution of sodium hypochlorite for one minute and then wash them in sterile water.

Place five embryo-containing pieces on one plate with the GA and five pieces without the embryo on the other GA plate. Repeat with the control plates without the GA. Incubate the plates at 25°C. for two days and then develop them by pouring on dilute iodine solution. Amylase activity is indicated by colourless areas where the starch has been hydrolysed. Deduce what you can about the effects of GA and the embryo on the activity of the enzyme amylase.

(Slightly modified after James, W. O., *An Introduction to Plant Physiology*, 7th edn. Clarendon Press, Oxford. Exp. 34, p. 165)

APPENDIX TO CHAPTER 8

1. Investigation of the effect of light on the germination of seeds.
2. Investigation of the effect of the phenolic substance coumarin on the germination of lettuce seeds
3. Investigation of the germination and growth of pollen by the hanging drop technique
4. Investigation of the effect of IAA on the abscission of leaf stalks.

1. Investigation of the effect of light on the germination of seeds
(see p. 180)

Material. Seeds of Love-in-a-Mist (*Nigella*). Some strains of lettuce e.g. Grand Rapids (obtainable from Thompson and Morgan (Ipswich Seeds Ltd.), Ipswich, Suffolk, U.K.). It is important to have fresh seeds. Some strains differ in their light requirements and also may lose their sensitivity on storage.

Method. Prepare four light-tight boxes as shown in fig. A.16. These are best made of wood and it is advisable to paint the inside black. Different qualities of red light can be obtained by the use of filters which should be fitted to two of the boxes.

Red light (about 650 nm). Cinemoid filter No. 14 (available from Rank Strand Electric Ltd. Order Processing Unit, Mitchelstone Estate, Kirkaldy, Fife,

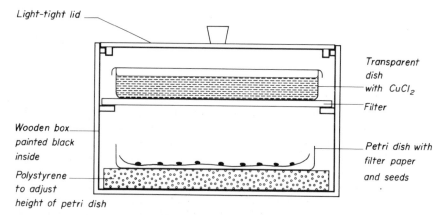

Fig. A.16. Box used in experiments on the light requirements for germination of seeds.

Scotland, U.K.). This should be used in conjuction with a filter made by placing a 1 per cent solution of $CuCl_2.2H_2O$ in a clear container if far red light is also to be removed (725 nm).

Far Red Light (about 725 nm). Cinemoid filter No. 14 together with cinemoid filter No. 20.

Place a thick filter paper in the bottom of each of four Petri dishes and moisten these with 5 cm³ distilled water. Place five dry seeds in each and immediately place the dishes in the light tight boxes with the lids in place. Leave the seeds in the dark for half an hour to imbibe water and then treat them with light (100 watt bulb 0.3 m above each plate), as follows:

Box A   Keep closed
Box B   No filter. Expose to white light for five minutes
Box C   Far red filters. Remove lid for five minutes
Box D   Red filter with $CuCl_2$. Remove lid and expose for five minutes.

Incubate the boxes at about 18°C. for three days, then open the boxes, remove the plates and score the number of seeds that have begun to germinate. If desired make allowance for the amount of light penetrating the various filters by varying the distance between the light and the boxes.

(After James, W. O., *An Introduction to Plant Physiology*, 7th edn, Clarendon Press, Oxford. Exp. 37, p. 166)

2. Investigation of the effect of the phenolic substance coumarin on the germination of lettuce seeds (see p. 174)

Material. Lettuce seeds (variety Grand Rapids).
Method. Take three Petri dishes and place a filter paper in the bottom of each. Place 10 cm³ distilled water in the first, 10 cm³ 40 parts per million (p.p.m.) coumarin solution in the next and 10 cm³ 20 p.p.m. coumarin solution in the third. Put 50 lettuce seeds in each dish, close the dishes and put them in a dark cupboard at about 25°C. Examine the seeds for germination, in diffuse light, every two days and record the germination percentage. Continue the investigation for about ten days. Calculate the percentage inhibition of lettuce germination caused by the coumarin solutions.

(After *Plant Hormones Set Booklet*, Experiment 3. Courtesy Philip Harris, Biological Ltd.)

3. Investigation of the germination and growth of pollen by the hanging drop technique (see p. 195)

Material. Ripe anthers from various plants, e.g. tulip (*Tulipa*), bluebell (*Endymion*), crocus, honeysuckle (*Lonicera*), plantain (*Plantago*).
Method (a) Prepare a 1.0 M nutrient solution by dissolving 171 g sucrose, 0.005 g boric acid and 0.05 g yeast in distilled water. Dilute this so as to provide solutions varying from 0.1 M to 0.5 M.
(b) Take a clean slide and place on it a polythene ring about 1 cm diameter (this can easily be made by cutting up polythene tubing). The ring should have been vaselined on both edges. Put a drop of the nutrient solution on the underside of a coverslip, remove an anther from the plant and dab this into the nutrient. Place

the coverslip, drop downwards, on the vaselined ring as shown in fig. 8.16. Place in a warm cupboard and examine with a high power microscope at hourly intervals. The method can be repeated for the range of solutions and for different plants. Determine which solution is the most suitable for the germination of pollen. Note the male nuclei and tube nucleus. It is sometimes possible to see cell division taking place in the pollen tube (see technique p. 241).

4. Investigation of the effect of IAA on the abscission of leaf stalks
(see p. 200)

Material. Coleus plants, with at least six pairs of leaves.
Method. Leaving the top three pairs of large leaves intact, carefully remove the leaf blade only of the next three pairs, leaving the leaf stalk or petiole about 1 cm long intact. Using a glass rod or syringe apply a small amount of lanolin paste containing IAA (see appendix p. 254) to three of the petioles at their cut ends. Alternatively the lanolin can be put into half a gelatine pill capsule and the whole placed over the cut petioles. Apply pure lanolin as a control to the other three. Keep the plants in the greenhouse for about three weeks and record when abscission of the leaf stalks takes place.

Label the stalks with paper reinforcement rings by placing these over the treated stalk.

(After *Plant Hormones Set Booklet*, Experiment 4. Courtesy Philip Harris, Biological Ltd.)

Appendix B. Useful Reagents

Aceto-lacmoid. One per cent solution dissolved in 45 per cent ethanoic (acetic) acid. Filter off undissolved solid after it has had some hours to dissolve. Add two drops of M-HCl to a watch-glassful of the stain before use. The stain deteriorates after about one month.

Aceto-orcein. One per cent solution dissolved in 45 per cent ethanoic (acetic) acid. Filter off undissolved solid after it has had some hours to dissolve. The stain is used without HCl and deteriorates after about one month.

Alcohol. See *Ethanol*.

Aniline sulphate. One per cent solution dissolved in distilled water. Filter and acidify with a few drops of dilute sulphuric acid.

Barium hydroxide water (0.95M solution). Dissolve about 50g of barium hydroxide and 15 g of barium chloride in 500 cm³ of boiling distilled water. Allow the solution to cool with a tube of soda-lime corked into the neck of the flask. The excess barium hydroxide will crystallize out and the clear fluid may be siphoned off into a second vessel, from which carbon dioxide has been removed, without allowing the solution to come into contact with the outer air. Add about a litre of freshly boiled distilled water, which has cooled under soda-lime, and standardize with 0.1 M-HCl, using phenol phthalein.

(From James: *An Introduction to Plant Physiology*, by permission of the Clarendon Press, Oxford.)

Benedict's reagent.

Solution A.: Weigh out 17.3 g copper II sulphate (hydrated) and dissolve this in 150 cm³ distilled water.

Solution B.: Weigh out 173 g sodium citrate (hydrated) and 90 g anhydrous sodium carbonate. Dissolve these in 850 cm³ distilled water and filter.

Add solution A to solution B slowly, with constant stirring. The reagent does not deteriorate on standing.

Borate buffer. Take 5 g pure boric acid and place it in a 500 cm³ flask. Add 100 cm³ absolute ethanol and 350 cm³ distilled water. Add 2.5 cm³ bromocresol green and 2.5 cm³ methyl red and adjust the pH to 5.0 (a reddish purple) using sodium hydroxide.

Brodie's fluid substitute. Add 22 g sodium bromide, 0.5 g stergene and 0.15 g Evans Blue to 500 cm³ distilled water and stir until dissolved.

Buffers. See borate buffer, phosphate buffer, sucrose buffers.

Chloral hydrate. Dissolve 160g of chloral hydrate in 100 cm³ distilled water.

Dinitrophenylhydrazine. Dissolve 0.1 g of 2:4-dinitrophenylhydrazine in dilute hydrochloric acid made by mixing 17 cm³ concentrated hydrochloric acid with 20 cm^3 of distilled water. Warm on a water-bath to dissolve the solid. Dilute the cold solution to 100 cm^3 with distilled water. The reagent is dilute and is used for the qualitative identification of substances containing aldehyde or ketone groups, with which it forms brownish crystal-line hydrazones. *Caution: dinitrophenylhydrazine is poisonous.*

Ethanol. Laboratory rectified spirit is about 95 per cent ethanol with 5 per cent methanol and is diluted as required.

Gibberellin (GA). A stock solution may be made up by dissolving 0.1 g Gibberellic acid in 2 cm^3 ethanol, this is then added to 100 cm^3 distilled water to give a solution of 1 part per 1,000. The solution may be kept in a dark bottle in the refrigerator for some weeks.

(After *Plant Hormones Set Booklet.* Courtesy Philip Harris, Biological Ltd.)

Guaiacum solution. Dissolve a few pieces of guaiacum resin in 95 per cent alcohol so as to make a pale coffee-coloured solution. Use the solution freshly prepared.

Indolylacetic acid (IAA). A stock solution may be made up by dissolving 0.1 g IAA in 2 cm^3 ethanol, this is then added to 900 cm^3 distilled water and warmed at 80°C. for 5 minutes. The volume is then made up to 1 dm^3. This stock solution (1 part per 10000) may be diluted as required and will keep in the refrigerator for some weeks.

(After *Plant Hormones Set Booklet.* Courtesy Philip Harris, Biological Ltd.)

Iodine. Dissolve 5 g of iodine crystals in 1 dm^3 of a strong solution of potassium iodide.

Lanolin with auxin. Dissolve 10 mg indolylacetic acid in a drop of ethanol and mix this thoroughly into 100 g hydrous lanolin. (If the lanolin is anhydrous, mix thoroughly with excess water, stand overnight, and decant the excess water.) The IAA-lanolin can then be conveniently stored in a plastic syringe and will keep in the refrigerator for two to three weeks if kept dark.

(After James, W. O., *An Introduction to Plant Physiology*, 7th edn. Clarendon Press, Oxford, p. 170)

Methyl green pyronin. Dissolve 1 g in 100 cm^3 distilled water. DNA stains green, RNA stains red.

Millon's reagent (Cole's version).

Solution A. Mix 100 cm^3 of 98 per cent sulphuric acid with 800 cm^3 distilled water and carefully add 100 g of mercury II sulphate, filter and make up to 1 dm^3. *This reagent should be handled with care.*

Solution B. 10 g of pure sodium nitrite dissolved in 1 dm^3 of distilled water.

Nadis' reagent. Can be obtained commercially and is made from three solutions:

Solution 1. Dimethyl-*p*-phenylene diamine solution.

Solution 2. 1-naphthol solution.

Solution 3. Sodium carbonate solution.

The reagent should be made up fresh using equal quantities of each solution.

Ninhydrin. Dissolve 0.2 g solid in 100 cm^3 butan-1-ol. It should be used freshly prepared. *This reagent should be handled with care.*

p-anisidine hydrochloride. Dissolve 1.5 g in 50 cm^3 butan-1-ol. The solution should be used freshly prepared.

Phenol phthalein. Dissolve 1 g in 100 cm³ ethanol.

Phenosafranin. Dissolve 1 g in 100 cm³ distilled water.

Phloroglucinol+HCl. Dissolve 1 g in 100cm³ ethanol and add a few drops of concentrated hydrochloric acid.

Phosphate buffer. Citric acid: Na_2HPO_4.pH6.5.
Citric acid solution 0.1 M. Dissolve 10 g citric acid in 500 cm³ distilled water.
Na_2HPO_4 *solution* 0.2 M. Dissolve 17.8 g Na_2HPO_4.$2H_2O$ in 500 cm³ distilled water.
Add 14.5 cm³ citric acid to 35.5 cm³ sodium phosphate. The buffer will then be about pH 6.5. Buffers at other pH values can be made using more or less of either reagent, but it is best to check the pH using a B.D.H. kit.

Pyrogallol. Take 10 g of sodium hydroxide and dissolve in 10 cm³ distilled water. Add this solution to one made by adding 2 g of pyrogallic acid to 6 g of distilled water. Stopper firmly with a rubber bung. *Take care with this highly caustic substance.*

Resorcinol+HCl. Dissolve 10 g resorcinol in 100 g acetone and add a few drops of concentrated HCl. The solution should be used freshly prepared.

Schultze's reagent. Dissolve 100 g of zinc in 300 cm³ of 31 per cent HCl. Evaporate the solution down to 150 cm³. During the evaporation a little extra zinc should be added. Take 12 g of potassium iodide and 0.15 g of iodine crystals and dissolve them in as little distilled water as possible. Mix the two solutions. If any precipitate forms the solution should be filtered through glass wool. Store in a dark, tightly stoppered bottle.

Sodium bicarbonate indicator

Preparation of stock solution. Dissolve 0.2 g of thymol blue and 0.1 g of cresol red in 20 cm³ ethanol. Add 0.84 g of pure sodium bicarbonate to 900 cm³ distilled water in a graduated flask. Add the dyes to this solution and make up to 1 dm³. Take care to exclude dust and dirt.

Preparation of final indicator. Pipette 25 cm³ of the stock solution into a 250 cm³ graduated flask and make up the volume with distilled water. The colour of this indicator should be red. If it is orange or yellow, aspirate atmospheric air through the solution.

(After *Biology Teachers Guide,* Year III, Nuffield Science Teaching Project. Longmans/Penguin, 1966)

Sodium cobaltinitrite. Dissolve 20 g cobalt nitrate and 30 g sodium nitrite in 75 cm³ dilute ethanoic (acetic) acid. This is made by diluting 10 cm³ glacial ethanoic acid with 75 cm³ distilled water. Blow air through the mixture until the evolution of nitrogen dioxide ceases. Filter the solution and dilute to 100 cm³ immediately before use.

(After Mansfield, T. A. (1970), *Stomata in new perspective,* S.S.R. no. 179, p. 316)

Sodium picrate papers. Make a saturated solution of picric acid (*Care*) by standing 100 cm³ distilled water over surplus picrate (2,4,6-trinitrophenol) crystals for a few days. Decant the saturated solution into a beaker and add sodium bicarbonate to neutralize the acid, until the effervescence ceases. It will generally be found that 0.5 g of bicarbonate is needed for 100 cm³ of saturated acid solution. Filter the sodium picrate solution into a shallow tray and soak filter papers evenly in this solution. Dry the papers by allowing evaporation to take place at normal temperatures away from strong sunlight. Cut the papers into strips about 5 ×50 mm and store in a tin or dark jar. The papers keep for several weeks and are satisfactory so long as bright yellow. *Caution : both picric acid and sodium picrate are poisonous ; crystals of picric acid are explosive.*

(From Nuffield Advanced Science (Biology) (1971), *Laboratory Book.* Penguin, Harmondsworth, p. 91)

Sucrose buffer for extracting chloroplasts

Mix: anhydrous disodium hydrogen phosphate (Na_2HPO_4)   2.8 g
potassium dihydrogen phosphate (KH_2PO_4)   6.4 g
sucrose   102.8 g
potassium chloride   37.0 g
distilled water to make up   1.0 dm^3

(After Nuffield Advanced Science (Biology) *Laboratory Book.* (1971). Penguin Harmondsworth, p. 76)

Sucrose buffer for extracting mitochondria

Mix: potassium dihydrogen phosphate (KH_2PO_4)   0.18 g
disodium hydrogen phosphate (Na_2HPO_4)   0.76 g
magnesium sulphate   0.10 g
sucrose   13.60 g
butanedioic acid (succinic acid)   1.36 g
sodium bicarbonate   1.68 g
distilled water to make up   100 cm^3

There is quite a vigorous reaction with the production of carbon dioxide when the succinic acid reacts with the sodium bicarbonate. Take care and finally stir and then shake the buffer until all the carbon dioxide is removed.

(After Nuffield Advanced Science (Biology) (1971). Laboratory Book. Penguin Harmondsworth, p. 75)

Appendix C. Bibliography of Practical Techniques

ABBOT, D. and ANDREWS, R. S. (1965). An Introduction to Chromatography. London, Longmans.

ALLEN, R. A., MILLETT, R. J. and SMITH, D. B. (1959). Radioisotope Data. London, H.M.S.O.

Autoradiography. Kodak Data Sheet SC-10.

CONWAY, E. J. (1950). Microdiffusion Analysis and Volumetric Error, 3rd edn. London, Crosby Lockwood.

FAIRES, R. A. and PARKS, B. H. (1958). Radioisotope Laboratory Techniques. London, George Newnes.

HALE, L. J. (1958). Biological Laboratory Data. London, Methuen.

JAMES, W. O. (1973). 7th edn. An Introduction to Plant Physiology. London, Clarendon Press.

LOOMIS, W. E. and SHULL, C. A. (1937). Experiments in Plant Physiology. New York, McGraw-Hill.

MCLEAN, R. C. and IVIMEY COOK, W. R. (1941). Plant Science Formulae. London, Longmans.

MEYER, B. S. and ANDERSON, D. B. (1955). Laboratory Plant Physiology, 3rd edn. Princeton, New Jersey, Van Nostrand.

MINER, H. A., et al. (1959). Teaching with Radioisotopes. Washington, U.S. Atomic Energy Commission.

Nuffield Advanced Science (Biology) (1971). Laboratory Guide. Penguin, Harmondsworth.

Oxoid Manual of Culture Media, The (1962). 2nd edn. Oxoid Division, Oxo Ltd., London S.E.1.

VOGEL, I. A. (1951). A Textbook of Practical Organic Chemistry, including Qualitative Organic Analysis. London, Longmans.

VOGEL, I. A. (1951). A Textbook of Quantitative Inorganic Analysis, 2nd edn. London, Longmans.

VOGEL, I. A. (1958). Elementary Practical Organic Chemistry. London, Longmans.

Index